From Breakthrough to Mainstream: How Turbochargers Revolutionized the Automobile

Georgia

Contents

6 Concluding remarks   175

Chapter 1

Introduction

Contents

1.1 Motivation

The invention of the automobile in Europe in the end of the 19th century, and its subsequent mass production in the early decades of the 20th century caused one of the greatest disruptions in the history of mankind. The newborn technology opened up a world of possibilities, providing not only greater individual freedom and autonomy, but also implied a new universe of opportunities in the industry, economic growth and in both domestic and foreign trade. From then onwards, thousands of millions of combustion engine-powered automobiles have been made around the world leading to a production of almost 100 million of automobiles only in the year 2017, resulting in a total of more than one billion of vehicles in circulation.

The emergence of forced-induction systems to the engines has become one of the most significant advancement in the automotive history. These systems allow to fill the cylinders with air at a higher pressure and density than using atmospheric pressure admission systems. This higher air density enables the introduction of larger amounts of fuel for generating the combustion and, in consequence, higher specific power and efficiency is achieved. Among the different types of forced-induction systems that have been developed along the history, the turbocharger system , composed by a radial turbine connected through a shaft with a centrifugal compressor, is by far the most extended one. In this system, the energy coming from the exhaust gases is recovered by the turbine and used to power the compressor that feeds the intake of the engine.

The invention of turbocompressors dates from the year 1905 and is attributed to the Swiss engineer Alfred Büchi, who worked in the research of diesel engines in an engine manufacturer company. A few years later, in the 1920s, engines equipped with turbochargers start appearing, mainly in the aeronautical field but also in the production of locomotives and ships. Due to the reduction of the effects of low atmospheric density, turbochargers constituted a huge step ahead in the engine performance of the aircraft of that time.

It is not until decades later, in the 1950s and 1960s, that the turbocharger is introduced in the automotive industry with the popularization of turbocharged large diesel engines in heavy duty applications such as trucks. However, at that time, turbocharger systems were not yet extended to the commercial passenger car mainly due to the low reliability with the highly variable loads typical of cars. Because of that, turbochargers start being widespread used in the 1980s in Compression Ignited engines (CI) and then, also extended to the Spark Ignited (SI) engines especially in the last decade.

In the last years, turbochargers have become a key factor for achieving better efficiency of the vehicle through a higher specific power, principally because it allowed a reduction of the cylinder displacement or even the necessary number of cylinders, a technique called "downsizing", a trend that has been massively extended during the last years.

In the last decades the immense economic growth of some Asian giants such as China or India has accelerated the demand in automobile production, reaching the current ratio of one automobile for every six world citizens. Needless to say that this figure will soon be updated, as analysts estimate that it is likely to end up doubling in 30 years. The emerging self-driving car technology might revert this trend in the long term, as it is supposed to reduce the total number of necessary vehicles by reducing the amount of time that each vehicle is stopped or in the garage. However, due to the actual immaturity of said technology, this is unlikely to take place in the following years.

In the near past, this never-ending rise in energy and transportation demands has materialised in a strong pressure in the environment and biosphere and even in an acceleration of the climate change [5]. Furthermore, several health issues derived from pollutant emission have appeared especially in the cities, leading to a growing concern on the part of the authorities and the society in general, and precipitating the increasingly stringent environmental regulations, as it will be shortly explained. In addition, although electrical engines are increasing its market share, internal combustion engines (ICE) have still a long way to go [6]. According to the data available [7] less than 0.5% of the global market of vehicles in circulation are electric. Thus, in the following decades it is likely that further development for making the ICE more efficient should be carried out while the transition to the electrical vehicle is completed.

Several measures and regulations against emission of pollutant gases have emerged in the recent years around the world. In 1993, the United Nations Framework Convection on Climate Change formulated a request for the nations to implement regional programmes containing measures to mitigate climate change. In that moment, the objective pursed by the European Union was a 30% reduction in greenhouse gas emissions by the year 2020, for the developed countries (compared to the 1990 levels). In 1998, the European Automobile Manufacturers Association adopted a commitment to reduce the average emissions from the new sold cars to the level of 140 g CO_2/km by 2009. In 1999, the Japanese and Korean Automobile Manufacturers Associations adopted a similar goal. Later on, in 2007, the European Comission admitted that, although significant progress had been made towards the maximum value of 140 g CO_2/km, the Community objective of averaged emissions of 120 g CO_2/km of new made cars would not be reached in the year of 2012. Since 2009, the European Commission sets mandatory emission reduction targets for new cars, with the initial objective of reaching the target of 130 g CO_2/km by 2015 and 95 g CO_2/km by 2020 [8]. This 2015 target was in fact reached in 2013, two years before the deadline. More recently, in 2018, the CO_2 levels for new cars registered in EU were 120.4 g CO_2/km [9], with an average emission reduction of 20 g CO_2/km (14.2%) since 2010. Finally, in a series of phases starting in 2020, the target from 2021 ahead will be an average emission of 95 g CO_2/km for new cars. In other big markets such as Japan or U.S.A and also China, similar regulations have been carried out during the recent years. But engine manufacturers are not only focused in CO_2 emissions, strict mandatory regulations also exists regarding the NO_x and soot emissions with regulations such as [10] and [11]. Apart from the pollutant emissions, noise emissions are an increasing concern for both regulators and consumers, being the second factor after pollution regarding the impact it has on health.

In this context of increasing concerns on the impact of pollutant emissions and highly stringent regulations, manufacturers and researchers have developed numerous technologies that have enabled the commercialization of more efficient engines in terms of fuel consumptions, and also cleaner in terms of pollutant emissions.

In particular, one of the main concerns is the emission of carbon dioxide (CO_2) that is liberated in the combustion of hydrocarbon fuels. The most immediate way of action to reduce CO_2 is to increase the global efficiency and specific power, i.e., reducing the fuel consumption while keeping or even increasing the engine performance. In this sense, as it has been explained in previous paragraphs, the most extended strategy is the aforementioned downsizing. Understanding the engine design from a downsizing perspective implies the reduction of engine size, namely, a reduction of the cylinder displacement and even the number of cylinders while incrementing the inlet pressure in the engine admission. In order to reach the necessary inlet pressure to obtain high engine performance with a smaller engine, a turbocharger is usually used, which allows to recover the energy from the exhaust gases to power the compressor placed in the intake line. Manufacturers are nowadays starting to slightly increase the engine displacement in a trend known as rightsizing, which nevertheless does not seem to be impacting a lot in the popularity of turbochargers. The turbocharger is a complex machine whose efficiency directly affects the whole engine efficiency. Because of that, a lot of efforts have been placed in improving the turbocharger technology:

- Two-stage boosting systems, which can allow a higher inlet pressure with a more optimised turbo range of operation as well as a reduction of the turbo lag and an improvement in the use of exhaust gas recirculation (EGR).

- Two-scroll turbines, which avoid interferences between exhaust processes of consecutive firing order cylinders and enable an optimization of scavenging effects and, in consequence, a higher volumetric efficiency. Also the transient response of the engine is improved due to a more effective use of dynamic pressure of the exhaust pressure pulses.

- Broad turbine and compressor optimisations, that include lowering the heat transfer and mechanical losses, improved stator blades and rotor geometry designs and better actuator and control systems among others.

On the other hand, not less important is the necessity to reduce the other main chemical subcomponents liberated in the different combustion of both, diesel or compression ignition (CI) engines and gasoline or spark ignition (SI) engines. In CI engines, the main contaminant agents are soot and NO_x. Soot emissions are mainly captured in particulate filters (DPF) and burnt later when the conditions in terms of accumulation and temperature are correct [12]. As for the NO_x emissions, which can damage the lung tissue and contribute to the formation of acid rain, are usually confronted by both active measures (avoiding their generation during the combustion) and passive measures (with aftertreatment technologies). In order to prevent the formation of NO_x it is crucial to reduce the maximum temperature reached in the cylinder during the combustion. To address this goal, the widely extended strategy is mixing part of the exhaust gas with the fresh air that is entering the admission, a technique called Exhaust Gas Recirculation (EGR). The main passive measure in this issue is the Selective Catalytic Reduction system (SCR), in which ammonia is injected into the exhaust gases stream reacting with the NO_x and resulting in N_2 and CO_2. For its part, in SI engines, the carbon monoxide (CO), the unburned hydrocarbons (HC) and nitrogen oxides (NO_x) are the main threatening pollutants which, fortunately, are drastically diminished by catalytic converters [13].

There are additional promising techniques for improving the efficiency of vehicles, that are currently being developed and that might be relevant in the following years, such as the rising usage of hybrid powertrain or the introduction of dual-fuel strategies. Turbochargers might be compatible with some of the future technologies that will arise in the following decades, nonetheless, what it is certain is that turbocharging improvements will be present in the immediate years to come, as it is expected that 50% of the cars sold will have turbochargers.

The turbocharger is a complex machine with multiple difficulties and challenges. Firstly, coupling the engine with a continuous flow turbocharger is traditionally a technical issue, especially due to the so called turbo-lag effect [14], understood as the delay between the moment in which the driver of the vehicle demands an increase of power, and the subsequent response of the turbocharger in order to reach the requested power. Secondly, the operating range in which modern downsized engines operate has forced the turbocharger to frequently work closer to surge line [15], i.e., combining high boosting pressure and low mass flow rates. Finally, with the completely extended downsizing trend and the new dynamic urban cycle regulations [16], the study of the pulsating flow has become crucial as they introduce several complex phenomena that impact engine performance. In particular, understanding the pulses in the turbine can lead to a more optimised design of the silencer and aftertreatment, as well as to a better prediction of the scavenging effects.

With the dramatical reduction in the computational cost, simulations and modelling and, in particular, one-dimensional codes have grown in importance in the last few years to the point that they constitute one of the main tools during engine design stage or during the development of an existing engine or technology. This scenario, along with the popularization of turbocharged and highly pulsating engines, drives us to a context in which accurate predictions of turbocharger performance under conditions of pulses of high amplitude and frequency will rise in relevance.

This **book** tries to contribute to the understanding of the aforementioned pulsating phenomena in turbochargers and to the modelling capacity in this topic. Firstly by exposing an experimental procedure for testing in highly pulsating conditions, then by presenting a model procedure, and finally with the explanation of a speed up process aimed to optimise the computational cost of the process of turbine acoustics simulation.

1.2 Background

During the last decades, engine modelling has been gaining prominence as a main procedure in the research and development field. As opposed to other traditional strategies as experiments, modelling and simulation campaigns are responsible for a massive reduction in time and costs associated with the design or improvement of engines and technologies. Manufacturers and researchers will continue to need experimental campaigns and trial-and-error-based procedures that complement the modelling and simulations and validate them, nevertheless, the majority of tests that traditionally were performed during the early design stages are nowadays replaced by numerical computations. When modelling, it will always exist a trade off between computational cost and model accuracy. Fortunately, the transistor density in dense integrated circuits doubles about every two years, incurring in proportional gains in the computational power. As a consequence, more detailed and complex model strategies have been feasible without increasing the computational cost. Keeping this in mind, it is always determinant to study the application and select the modelling approach in consequence, as computational cost is still important, especially when applied to fast simulation tools with real time capabilities.

There are different computational methods according to the level of complexity and similarity to the real physics. Most of them can be classified as:

- Look-up table-based methods. These methods require a large database in order to generate the outputs of the problem, and they usually need the experience of the engineers in order to interpret and configure them. This is probably the fastest way to obtain a first model due to its simplicity, but it has low accuracy and low time-resolution. Thus, they are mainly used in the early stages of engine design or for specific application.

- Mean value engine models. They do not produce instantaneous results, instead, they are mainly used for obtaining cycle averaged results. Mean value procedures are not as fast as a look-up tables but they can still be used for real time applications such as hardware in the loop (HIL).

- Filling and emptying models. In this approach, the momentum conservation between elements is modelled. They have time resolutions of the order of the crank angle so it is not always possible to use them in real time applications due to the higher computational cost. With this procedure, the capacity to predict non-linear behaviour is limited to the low-frequency range.

- One dimensional models. In this modelling approach, the momentum conservation equation is simulated and a better time resolution is feasible. Thus, these methods are used for modelling more complex systems, in which the non-linear acoustics performance is determinant as they are able to predict the wave dynamic behaviour.

- Quasi-2D methods. The philosophy of these methods is to introduce some improvements when comparing to pure one-dimensional models for solving a particular problem but without implying a much higher computational cost. The simplification adopted will depend on the specific problem that need to be solved, but the objective is to improve the performance of a classic one-dimensional model without the computational cost that of more detailed approaches.

- Three-dimensional methods, also known as computational fluid dynamic methods (CFD). Different mathematical Navier-Stokes based approaches are used to model as much phenomena as possible as possible, including Reynolds-averaged Navier-Stokes equations (RANS) solvers and turbulence-resolved simulation such as Large Eddy Simulations (LES) and Detached-Eddy Simulations (DES). The large increase of computational power of the last years has enable to simulate even the largest turbulent structures for some applications. Nevertheless, the computational power available is currently not enough to use these methods in the early and middle stages of engine design and thus, they are strictly used only where a high grade of detail is necessary.

Deepening into the turbocharger modelling, it can be observed how a wide variety of methods have been used with different level of complexity, with the most simple approach being a simple look-up table based in the manufacturer's maps. Look-up table models essentially use several tuning parameters to correct the turbocharger variables (efficiency, mass flow, etc) in order to take into consideration the effect that the mechanical losses, heat transfer or unsteady behaviour generate in conditions different from the ones found during the measurements of the maps. In the past, finding these corrections in order to correctly predict the turbocharger behaviour under this variety of conditions was feasible. However, modern turbochargers are smaller and they work in conditions in which these effects are gaining a lot of importance, specially in specific exigent engine situations such as an engine warm up or a real urban driving cycle, increasingly present in actual and future regulations. In this context, notable efforts have been made to develop one-dimensional codes more accurate and reliable while keeping the computational cost still very low. Furthermore, one-dimensional codes are crucial as the non-linear acoustics are becoming more and more important and acoustic pollution regulations more strict.

The work presented in this book can be considered a contribution for the process of continuous evolution of the complete one-dimensional wave action modelling soft-ware developed in CMT-Motores Térmicos in Universitat Politècnica de València [17]. This software was first initiated by Corberán [18] based on the postulates exposed by Benson [19] and making use of the method of characteristics as the procedure to solve hyperbolic partial differential equations systems. Years later, the method of characteristics was substituted by a faster and more conservative finite difference method by means of a second order two-step Lax-Wendroff scheme. Nowadays, the evolution of the flow inside the ducts is computed by means of a Finite Volume Method with a second order Monotone Upstream-centered Scheme for Conservation (MUSCL [20]) and the Harten-Lax-van Leer-Contact approximation for Riemann problem solving [21].

Several sub-models have been added to the OpenWAM software for different engine components such as filling and emptying models for the cylinders and plenums [18, 22] or the turbine boundary conditions [23], although here the focus is placed in the turbocharger sub-models. In this sense, numerous improvements have been tested and implemented in the OpenWAM in-house built software in a dynamic of constant evolution for the past years. With the fundamentals extracted from [24], the basis of one-dimensional turbocharger model implemented in the code is developed in [25] and further developed in [26]. The modularity of the code allows to easily incorporate new sub-models that solve complex problems such as the heat transfer in the turbocharger [27, 28, 29] the turbine map extrapolation [30] and compressor map extrapolation [31], or the mechanical losses [32]. Considering the work developed in this book as a framework, new contributions to the OpenWAM software have been implemented, as the quasi-2D turbine model [1, 33] exposed in chapter 2 or the two-scroll turbine model [3, 34] in chapter 4. Also the fast acoustic turbine model first presented in [4] and detailed in chapter 5 is currently under implementation.

This work is particularly focused on modelling and simulations of single and two-scroll turbines in non-steady pulsating conditions, where some other research teams have made meaningful contributions, such is the case of the Imperial College team in London. Costall et al. documented some findings related to the unsteady behaviour of twin entry turbines under full and unequal admissions [35, 36] and Romagnolli et al. for steady conditions [37]. The university of Malaysia also with collaboration of Imperial College made important investigations in this regard. In [38] Chiong et al. focus in the prediction of the twin-entry turbine performance in an application orientated approach, while in [39] incorporates the consideration of unequal and partial admission conditions. Finally, in [40] Chiong et al. exposed a one-dimensional method to predict the twin-scroll turbine performance under pulses that is furthermore validated against experimental data. In addition, in [41], an interesting comparison between single-scroll and twin-scroll turbines performance is presented for the Worldwide harmonized Light vehicles Test Cycle (WLTC). Collaborating with the same research group, Rajoo et al. made some interesting contributions, as the experimental study presented in [42] for characterising the turbine behaviour under a given pulse, analysing different vane angles as it was previously exposed in [43]. Other studies were performed for analysing the twin and double entry turbines under pulses, as the investigations carried out by Copeland et al. [44, 45].

The research institute CMT-Motores Térmicos of Universitat Politècnica de València has a long and prolific background in the field of turbocharger research and one-dimensional ICE code development. In this sense, several dissertations have preceded the contributions carried out by the respondent and have been essential for the materialization of this document. Using the first studies by Payri [46] and later Serrano [47] as starting point, numerous dissertations have been developed during the last two decades. Cervelló [48] developed a testing methodology for turbochargers and also made contributions to the variable geometry turbine modelling. Soon after that, in 2008, Tiseira Izaguirre [49] faced the compressor surge problem from both modelling and experimen-tal points of view. The PhD book by Fajardo [50] presented an exhaustive research of pulsating flow characterisation on radial turbines by means of computational fluid dynamics (CFD). Navarro [51] contributed with a method for predicting flow-induced acoustics, whereas Reyes-Belmonte [52] presented a very complete book in which, apart from one-dimensional modelling, he developed a procedure to account for the heat transfer losses in turbochargers. In the work by García-Cuevas [53] an innovative model for turbine simulation with a quasi-2D approach is presented, along with a new procedure for modelling the mechanical losses of the whole turbocharger. More recently, Artem Dombrovsky [54] described an improved model able to extrapolate the radial turbine performance in terms of VGT position, rotational speed and blade speed ratio. Later, Inhestern [55] added the study of the turbine performance under extreme off-design conditions . Regarding the compressor, Daniel Tarí [56] focused on the condensation problem and the influence of compressor geometry by means of both experiments and 3D-CFD simulations. In his PhD book, Manuel Hernández [57] exposed an innovative quasi-3D model for simulating non-linear conditions in air management elements such as a exhaust systems. Also recently, García Tíscar [58] presented a methodology for noise and acoustics measurements in compressors along with CFD simulations. The institute is now expanding its research in the area of two-scroll turbocharger systems as with PhD book by Vishnu Samala [59] which will be defended this year.

1.3   Objectives

The general objective of the present work is to contribute to the understanding and the prediction capabilities of radial turbines by developing new models and methods as well as improving existing models from the literature. These models are aimed to predict the turbocharger performance not only under standard characterisation conditions but also under highly pulsating and close to similar real driving conditions. Furthermore, to ensure model reliability and accuracy, models and methods developed should be fully validated by experimental campaigns. Thus, the challenges that the present work seeks to solve fall into two main categories, i.e., development of experimental procedures, and modelling work contributions.

As it has been summarised in section 1.2, several methods for testing single entry turbines under both steady and pulsating conditions can be found in the literature.

Also fewer but instead relevant works have been developed in the particular field of experimental testing of two-scroll turbines. In this regard, the present work tries to contribute to the state of the art with the following objectives:

- Develop an experimental procedure to test two-scroll turbines under simulated real driving conditions that is flexible enough to be used in a wide variety of pulsating flow conditions with minimal changes in the installation. This includes hot and cold flow and high or low pulse amplitude and mass flow.

- Provide a testing methodology that is easily adaptable to any type of two-scroll turbocharger.

- The design of the test rig ensures that the amplitude of the pulses fed to the turbine are high enough to enter the non-linear domain.

Regarding the modelling category, the major milestones to be achieved are the following:

- On the basis of the quasi-bidimensional turbine model first presented in [53], the new features added that will be described in (chapter 2) are intended to improve the accuracy of the model without adding significant computational cost, with the aim of providing an useful tool to the industry for specific applications.

- Develop an integrated two-scroll turbine model able to be used under unsteady conditions and aimed to predict the non-linear phenomena of highly pulsating flow even in the higher than 1000 Hz range.

- Provide the two-scroll turbine model with the necessary features of compatibility so it can be utilized along with the rest of submodels from the literature such as heat transfer models or map extrapolation models.

- Build a fast-running method that translates the modelling work developed in this book into a fast applicable tool for real time simulations.

- Additionally, the modelling work of the present research aims to deliver full compatibility with the de-facto industry standard modelling software GT-POWER and Ricardo WAVE.

1.4   Employed method

Turbochargers in general, and radial turbines in particular, are complex machines in which several phenomena take place at the same time such as complex fluid dynamics, heat transfer or mechanical losses. This work focuses in the study of the non-linear pulsating effects in the turbine with the following procedure as a guide:

- Study of the performance under pulsating flow conditions in a turbocharger gas stand. The methodology used needs to be well defined and reliable in terms of both procedures and instrumentation selected. In this regard, the wide in-house experience has been crucial for extracting high quality data and establishing well documented experimental procedures [2].

- For the modelling and validation work, a from-the-simple-to-the-complex approach has been firmly adopted. New features and modifications to the models were first implemented in a virtual gas stand and, once validated in the gas stand, then added to the complete engine model software. In the same way, unsteady conditions introduce a lot of unpredicted results and potential code errors, because of that, steady conditions were used as a first test to find potential problems prior to fully validate the model in pulsating conditions.

- The one dimensional code in which the models are implemented is OpenWAM [17], the 1D simulation integrated tool developed in CMT-Motores Térmicos. Later, the piece of software correspondent to the turbocharger models of this work have been extracted and coupled as a plug-in to the well known one-dimensional commercial software Ricardo WAVE and GT-POWER.

- Once the models are validated, the effort is then relocated into minimizing the computational cost by exploring new solutions that enable to apply the developed models in fast or real time applications while keeping as much as possible of the model accuracy.

1.5 book outline

After the introduction of the present chapter, the subsequent chapters are organised in the following manner:

Chapter 2 includes the last update of the quasi-bidimensional radial turbine model first presented in [53] and the corresponding experimental validation. The main contributions to the topic developed in this work are exposed along with the latest results. The experiments used for the validation of the model are also summarised, with the aim of being helpful to understand the methodology of the novel experimental method later explained in chapter 3.

The description of the experimental work carried out for successfully elaborate this book is concentrated in chapter 3. There, an innovative technique for testing two-scroll turbines under non-linear pulsating conditions is presented. First, the experimental facility and the procedure are described in detail. Then, an analysis of the results is performed in order to demonstrate the potential of the method for evaluating the performance of two-scroll turbines under simulated realistic conditions.

A complete integrated 1D twin-scroll turbine model able to be used in reciprocating internal combustion engine unsteady simulations is presented in chapter 4. The big amount of data collected in the experiments described in chapter 3 are here used to perform the experimental validation. Thus, it is shown how the model is able to correctly predict the two-scroll turbine performance under a wide variety of non-linear pulsating conditions.

After having presented the experiments in chapter 3 and the modelling work in chapters 2 and 4, a method for estimating the turbine acoustics with low computational cost is introduced in chapter 5. The turbine models developed along the previous chapters are used to feed the look-up tables that enable to predict the sound and noise in real-time simulations by means of acoustic transfer matrices.

Finally, chapter 6 summarises the concluding remarks of the book, including the main contributions presented in this work related to different topics, the conclusions obtained from the results of each of the main work sections, and also some recommen-dations for future studies on these regards.

Chapter 1 References

[1]   J. Galindo, F. J. Arnau, L. M. García-Cuevas, and P. Soler. "Experimental validation of a quasi-two-dimensional radial turbine model". *International Journal of Engine Research* (2018). ISSN: 1468-0874. DOI: 10.1177/1468087418788502 (cit. on pp. xi, 8, 57, 62, 84, 90, 113, 131, 133, 139, 177).

[2]   J. Serrano, F. Arnau, L. M. García-Cuevas, P. Soler, L. Smith, R. Cheung, and B. Pla. "An Experimental Method to Test Twin and Double Entry Automotive Turbines in Realistic Engine Pulse Conditions". In: *WCX SAE World Congress Experience*. SAE Technical Paper 2019-01-0319. 2019. DOI: 10.4271/2019-01-0319 (cit. on pp. xi, 11, 100, 130).

[3]   J. R. Serrano, F. J. Arnau, L. M. García-Cuevas, P. Soler, and R. Cheung. "Experimental validation of a one-dimensional twin-entry radial turbine model under non-linear pulse conditions". *International Journal of Engine Research* (2019), p. 146808741986915. ISSN: 1468-0874. DOI: 10.1177/1468087419869157 (cit. on pp. xi, 8).

[4]   A. Torregrosa, L. M. García-Cuevas, L. B. Inhestern, and P. Soler. "Radial turbine sound and noise characterisation with acoustic transfer matrices by means of fast one-dimensional models". *International Journal of Engine Research* (2019), p. 1468087419889429. DOI: 10.1177/1468087419889429. eprint: https://doi.org/10.1177/1468087419889429 (cit. on pp. xi, 8).

[5]   J. S. Gaffney and N. A. Marley. "The impacts of combustion emissions on air quality and climate - From coal to biofuels and beyond". *Atmospheric Environ-*

ment 43(1) (2009). Atmospheric Environment - Fifty Years of Endeavour, pp. 23 –36. ISSN: 1352-2310. DOI: 10.1016/j.atmosenv.2008.09.016 (cit. on p. 3).

[6] J. Serrano. "Imagining the Future of the Internal Combustion Engine for Ground Transport in the Current Context". *Applied Sciences* 7(10) (2017), p. 1001. ISSN: 2076-3417. DOI: 10.3390/app7101001 (cit. on pp. 3, 56).

[7] P. IEA. *Global EV Outlook 2019*. 2019. URL: www.iea.org/publications/ reports/globalevoutlook2019/ (cit. on p. 3).

[8] European Parliament and Council of the European Union. "Regulation (EU) No 510/2011 of the European Parliament and of the Council of 11 May 2011 setting emission performance standards for new light commercial vehicles as part of the Union's integrated approach to reduce CO 2 emissions from light-duty vehicles (Text with EEA relevance)". In: *Official Journal of the European Union 54*. 2011. DOI: 10.3000/17252555.L_2011.145.eng (cit. on p. 3).

[9] C. EEA. *Monitoring of CO2 emissions from passenger cars – Regulation (EC) No 443/2009*. 2019. URL: https://www.eea.europa.eu/data-and-maps/data/ co2-cars-emission-16 (cit. on p. 3).

[10] European Parliament and Council of the European Union. "Regulation (EC) No 715/2007 of the European Parliament and of the Council of 20 June 2007 on type approval of motor vehicles with respect to emissions from light passenger and commercial vehicles (Euro 5 and Euro 6) and on access to vehicle repair and maintenance information (Text with EEA relevance)". In: *Official Journal of the European Union 50*. 2007. DOI: 10.3000/17252555.L_2011.145.eng (cit. on p. 3).

[11] European Parliament and Council of the European Union. "Regulation (EC) No 595/2009 of the European Parliament and of the Council of 18 June 2009 on type-approval of motor vehicles and engines with respect to emissions from heavy duty vehicles (Euro VI) and on access to vehicle repair and maintenance information and amending Regulation (EC) No 715/2007 and Directive 2007/46/EC and repealing Directives 80/1269/EEC, 2005/55/EC and 2005/78/EC (Text with EEA relevance)". In: *Official Journal of the European Union 52*. 2009. DOI: 10.3000/17252555.L_2011.145.eng (cit. on p. 3).

[12] J. R. Serrano, H. Climent, P. Piqueras, and E. Angiolini. "Filtration modelling in wall-flow particulate filters of low soot penetration thickness". *Energy* 112 (2016), pp. 883 –898. ISSN: 0360-5442. DOI: 10.1016/j.energy.2016.06.121 (cit. on p. 4).

[13] P. Michel, A. Charlet, G. Colin, Y. Chamaillard, G. Bloch, and C. Nouillant. "Optimizing fuel consumption and pollutant emissions of gasoline-HEV with catalytic converter". *Control Engineering Practice* 61 (2017), pp. 198 –205. ISSN: 0967-0661. DOI: 10.1016/j.conengprac.2015.12.010 (cit. on p. 5).

[14]   F. Millo, F. Mallamo, E. Pautasso, and G. Ganio Mego. "The Potential of Electric Exhaust Gas Turbocharging for HD Diesel Engines". In: *SAE 2006 World Congress & Exhibition*. SAE International, 2006. DOI: 10.4271/2006-01-0437 (cit. on p. 5).

[15]   D. Evans and A. Ward. "Minimizing Turbocharger Whoosh Noise for Diesel Powertrains". *SAE Technical Paper* 2005-01-2485 (2005). DOI: 10.4271/2005-1-2485 (cit. on p. 5).

[16]   B. WLTP facts. *What is WLTP and how does it work*. 2019. URL: https://wltpfacts.eu/what-is-wltp-how-will-it-work/ (cit. on p. 5).

[17]   CMT – Motores T'ermicos, Universitat Politècnica de València. *OpenWAM*. 2016. URL: http://www.openwam.org/ (cit. on pp. 7, 11, 22, 98, 133, 178).

[18]   J. M. Corberán. *Contribución al modelado del proceso de renovación de la carga en motores de combustión interna alternativos*. PHD book. 1984 (cit. on pp. 7, 8).

[19]   R. Benson. "The thermodynamics and gas dynamics of internal-combustion engines". In: *Proceedings of the IEEE*. Vol. 66. IEEE, 1978, pp. 51–83. DOI: 10.1109/PROC.1978.10837 (cit. on p. 7).

[20]   B van Leer. "Towards the ultimate conservative difference scheme, V. A second order sequel to Godunov's method". *Journal of Computational Physics* 32 (1979), pp. 101–136. DOI: 10.1016/0021-9991(79)90145-1 (cit. on p. 8).

[21]   E. Toro, M. Spruce, and W. Speares. "Restoration of the contact surface in the HLL-Riemann solver". English. *Shock Waves* 4(1) (1994), pp. 25–34. ISSN: 0938-1287. DOI: 10.1007/BF01414629 (cit. on pp. 8, 23).

[22]   F. Payri, J. Desantes, and J. Corberán. "A quasi-steady model on gas exchange process, some results". In: *Motor Sympo 88*. 1988 (cit. on p. 8).

[23]   F. Payri, J. Benajes, and M. Reyes. "Modelling of supercharger turbines in internal-combustion engines". *International Journal of Mechanical Sciences* 38(8) (1996), pp. 853 –869. ISSN: 0020-7403. DOI: 10.1016/0020-7403(95)00105-0 (cit. on pp. 8, 91).

[24]   H. Chen and D. Winterbone. "A method to predict performance of vaneless radial turbine under steady and unsteady flow conditions". In: *Turbocharging and Turbochargers*. Institution of Mechanical Engineers. 1990, pp. 13–22 (cit. on pp. 8, 20, 21, 97).

[25]   J. R. Serrano, F. J. Arnau, V. Dolz, A. Tiseira, and C. Cervelló. "A model of turbocharger radial turbines appropriate to be used in zero- and one-dimensional gas dynamics codes for internal combustion engines modelling". *Energy Conversion and Management* 49(12) (2008), pp. 3729 –3745. ISSN: 0196-8904. DOI: 10.1016/j.enconman.2008.06.031 (cit. on pp. 8, 30, 31, 96, 99, 100, 134).

[26] J. R. Serrano, F. J. Arnau, R. Novella, and M. Á. Reyes-Belmonte. "A Procedure to Achieve 1D Predictive Modeling of Turbochargers under Hot and Pulsating Flow Conditions at the Turbine Inlet". *SAE Technical Paper* 2014-01-1080 (2014), 13pp. DOI: 10.4271/2014-01-1080 (cit. on p. 8).

[27] F. Payri, P. Olmeda, F. J. Arnau, A. Dombrovsky, and L. Smith. "External heat losses in small turbochargers: Model and experiments". *Energy* 71 (2014), pp. 534 –546. ISSN: 0360-5442. DOI: 10.1016/j.energy.2014.04.096 (cit. on pp. 8, 20, 27, 92, 131).

[28] J. R. Serrano, P. Olmeda, F. J. Arnau, A. Dombrovsky, and L. Smith. "Turbocharger heat transfer and mechanical losses influence in predicting engines performance by using one-dimensional simulation codes". *Energy* 86 (2015), pp. 204 –218. DOI: 10.1016/j.energy.2015.03.130 (cit. on pp. 8, 27, 92, 131).

[29] A. Gil, A. Tiseira, L. M. García-Cuevas, T. Rodríguez Usaquén, and G. Mijotte. "Fast three-dimensional heat transfer model for computing internal temperatures in the bearing housing of automotive turbochargers". *International Journal of Engine Research* (2018). DOI: 10.1177/1468087418804949 (cit. on pp. 8, 27, 92, 131).

[30] J. R. Serrano, F. J. Arnau, L. M. García-Cuevas, A. Dombrovsky, and H. Tartoussi. "Development and validation of a radial turbine efficiency and mass flow model at design and off-design conditions". *Energy Conversion and Management* 128 (2016), pp. 281 –293. ISSN: 0196-8904. DOI: 10.1016/j.enconman.2016.09.032 (cit. on pp. 8, 26–29, 96, 98, 131, 134, 135, 139, 157, 178).

[31] J. Galindo, R. Navarro, L. M. García-Cuevas, D. Tarí, H. Tartoussi, and S. Guilain. "A zonal approach for estimating pressure ratio at compressor extreme off-design conditions". *International Journal of Engine Research* 20(4) (2019), pp. 393–404. DOI: 10.1177/1468087418754899 (cit. on pp. 8, 92, 131).

[32] J. R. Serrano, P. Olmeda, A. Tiseira, L. M. García-Cuevas, and A. Lefebvre. "Theoretical and experimental study of mechanical losses in automotive turbochargers". *Energy* 55(0) (2013), pp. 888 –898. ISSN: 0360-5442. DOI: 10.1016/j.energy.2013.04.042 (cit. on pp. 8, 21, 92, 131).

[33] J. Galindo, H. Climent, A. Tiseira, and L. M. García-Cuevas. "Effect of the numerical scheme resolution on quasi-2D simulation of an automotive radial turbine under highly pulsating flow". *Journal of Computational and Applied Mathematics* 291 (Jan. 2016), pp. 112–126. DOI: 10.1016/j.cam.2015.02.025 (cit. on pp. 8, 131).

[34] J. Serrano, F. Arnau, L. Garcéa-Cuevas, and V. Samala. "Development of flow oriented model for extrapolation and interpolation off-design performance of twin-entry and dual volute radial-inflow turbines working under different flow admission conditions". *Energy* (). Under review (cit. on pp. 8, 92, 99, 100, 177).

[35] A. W. Costall, R. M. McDavid, R. F. Martínez-Botas, and N. C. Baines. "Pulse performance modelling of a twin-entry turbocharger turbine under full unequal admission". In: *Proceedings of ASME Turbo Expo 2009*. 2009. ASME, 2009. DOI: 10.1115/1.4000566 (cit. on pp. 8, 20, 130).

[36] A. W. Costall, R. M. McDavid, R. F. Martinez-Botas, and N. C. Baines. "Pulse Performance Modeling of a Twin Entry Turbocharger Turbine Under Full and Unequal Admission". *Journal of Turbomachinery* 133(2) (2011), p. 021005. ISSN: 0889504X. DOI: 10.1115/1.4000566 (cit. on pp. 8, 91).

[37] A. Romagnoli, C. D. Copeland, R. Martinez-Botas, M. Seiler, S. Rajoo, and A. Costall. "Comparison Between the Steady Performance of Double-Entry and Twin-Entry Turbocharger Turbines". *Journal of Turbomachinery* 135(1) (2013). ISSN: 0889-504X. DOI: 10.1115/1.4006566 (cit. on pp. 8, 56).

[38] M. S. Chiong, S. Rajoo, R. F. Martinez-Botas, and A. W. Costall. "Engine turbocharger performance prediction: One-dimensional modeling of a twin entry turbine". *Energy Conversion and Management* 57 (2012), pp. 68–78. ISSN: 01968904. DOI: 10.1016/j.enconman.2011.12.001 (cit. on pp. 8, 91).

[39] M. S. Chiong, S. Rajoo, A. Romagnoli, A. W. Costall, and R. F. Martinez-Botas. "Assessment of Partial-Admission Characteristics in Twin-Entry Turbine Pulse Performance Modelling". In: *Volume 2C: Turbomachinery*. 2015, V02CT42A022. ISBN: 978-0-7918-5665-9. DOI: 10.1115/GT2015-42687 (cit. on pp. 8, 24, 92).

[40] M. S. Chiong, S. Rajoo, A. Romagnoli, A. W. Costall, and R. F. Martinez-Botas. "One-dimensional pulse-flow modeling of a twin-scroll turbine". *Energy* 115 (2016), pp. 1291–1304. ISSN: 03605442. DOI: 10.1016/j.energy.2016.09.041 (cit. on p. 8).

[41] M. S. Chiong, M. A. Abas, F. X. Tan, S. Rajoo, R. Martinez-Botas, Y. Fujita, T. Yokoyama, S. Ibaraki, and M. Ebisu. "Steady-State, Transient and WLTC Drive-Cycle Experimental Performance Comparison between Single-Scroll and Twin-Scroll Turbocharger Turbine". In: *WCX SAE World Congress Experience*. SAE International, 2019. DOI: https://doi.org/10.4271/2019-01-0327 (cit. on pp. 8, 56).

[42] S. Rajoo, A. Romagnoli, and R. F. Martinez-Botas. "Unsteady performance analysis of a twin-entry variable geometry turbocharger turbine". *Energy* 38(1) (2012), pp. 176–189. ISSN: 03605442. DOI: 10.1016/j.energy.2011.12.017 (cit. on pp. 8, 57, 91).

[43] S. Rajoo and R. F. Martinez-Botas. "Variable Geometry Mixed Flow Turbine for Turbochargers: An Experimental Study". *International Journal of Fluid Machinery and Systems* 1(1) (2008), pp. 155–168. ISSN: 1882-9554. DOI: 10.5293/IJFMS.2008.1.1.155 (cit. on pp. 8, 91).

[44]   C. D. Copeland, R. F. Martinez-Botas, and M. Seiler. "Comparison Between Steady and Unsteady Double-Entry Turbine Performance Using the Quasi-Steady Assumption". *Journal of Turbomachinery* 133(3) (2011), p. 031001. ISSN: 0889504X. DOI: 10.1115/1.4000580 (cit. on p. 8).

[45]   C. D. Copeland, R. F. Martinez-Botas, and M. Seiler. "Unsteady Performance of a Double Entry Turbocharger Turbine With a Comparison to Steady Flow Conditions". *Journal of Turbomachinery* 134(2) (2012), p. 021022. ISSN: 0889504X. DOI: 10.1115/1.4003171 (cit. on pp. 8, 91).

[46]   F. Payri. "Predicción de las actuaciones de los grupos de sobrealimentación para motores diesel de automoción". PhD book. Universitat Politècnica de València, 1973 (cit. on p. 9).

[47]   J. R. Serrano. "Análisis y modelado de transitorios de carga en MEC turboalimentados". PhD book. Universitat Politècnica de València, 1999 (cit. on p. 9).

[48]   C. Cervelló. "Contribución a la Caracterización Experimental y al Modelado de Turbinas de Geometría Variable en Grupos de Sobrealimentación". PhD book. Universitat Politècnica de València, 2005 (cit. on p. 9).

[49]   A. Tiseira. "Caracterización experimental y modelado de bombeo en compresores centrífugos de sobrealimentación". PhD book. Universitat Politècnica de València, 2008 (cit. on p. 9).

[50]   P. Fajardo. "Methodology for the Numerical Characterization of a Radial Turbine under Steady and Pulsating Flow". PhD book. Universitat Politècnica de València, 2012 (cit. on p. 9).

[51]   R. Navarro. "A numerical approach for predicting flow-induced acoustics at near-stall conditions in an automotive turbocharger compressor". PhD book. Univer-sitat Politècnica de València, 2014 (cit. on p. 9).

[52]   M. Ángel Reyes-Belmonte. "Contribution to the Experimental Characterization and 1-D Modelling of Turbochargers for IC Engines". PhD book. Universitat Politècnica de València, 2013 (cit. on p. 9).

[53]   L. M. G.-C. González. "Experiments and Modelling of Automotive Turbochargers under Unsteady Conditions". PhD book. Universitat Politècnica de València, 2014 (cit. on pp. 9–11, 176–178).

[54]   A. Dombrovsky. "Synthesis of the 1D modelling of turbochargers and its effects on engine performance prediction". PhD book. Universitat Politècnica de València, 2016 (cit. on p. 9).

[55]   L. Inhestern. "Measurement, Simulation, and 1D-Modeling of Turbocharger Ra-dial Turbines at Design and Extreme Off-Design Conditions". PhD book. Uni-versitat Politècnica de València, 2014 (cit. on p. 9).

[56]   D. Tarí. "Effect of inlet configuration on the performance and durability of an automotive turbocharger compressor". PhD book. Universitat Politècnica de València, 2018 (cit. on p. 9).

[57] M. Hernández. "A non-linear quasi-3D model for air management modelling in engines". PhD book. Universitat Politècnica de València, 2018 (cit. on p. 9).

[58] J. García-Tíscar. "Experiments on Turbocharger Compressor Acoustics". PhD book. Universitat Politècnica de València, 2017. URL: http://hdl.handle.net/10251/79552 (cit. on p. 9).

[59] V. Samala. "Experimental characterization and mean line modelling of twin-entry and dual-volute turbines working under steady with different flow admission conditions". PhD book. Universitat Politècnica de València, 2020 (cit. on p. 9).

Chapter 2

Contributions to a quasi-two-dimensional radial turbine model

Contents

2.1 Introduction

Roughly one third of the energy released during the combustion phase in an reciprocating internal combustion engine (ICE) escapes the cylinder and is available in the exhaust manifold to be recovered [60]. This high pressure and temperature gas can be expanded in a turbine to drive a compressor, leading to higher engine power densities and lower fuel consumption and emissions. However, the application of this technology requires the study of coupling strategies between the engine and the turbocharger, being the variable geometry turbine (VGT) the most extended solution in diesel applications. Although it is difficult to use a VGT with a petrol engine due to the high exhaust temperatures, it is currently being evaluated as a very attractive way of driving down emissions and fuel consumption also in these engines, as described in [61].

Most of the fluid-dynamic processes that take place in the engine can be reproduced by means of zero-dimensional and simple one-dimensional models, providing reasonable precision with very low computational cost. In this context, using mainly the characteristic curves could be the most efficient solution. Nevertheless, that implies using information of previous measurements to characterise the curves in an operating range that is usually wider that the one provided by manufacturers. In addition, it makes mandatory to interpolate and makes impossible to extrapolate or to consider mass accumulation during unsteady operation.

For that reason, the correct physical modelling of the variable geometry turbine, with an evolved one-dimensional model, provides a powerful tool for the analysis and development of reciprocating engines, and in particular, valuable information for the design of the necessary matching between engine and turbocharger plus the required control strategies. Active control systems such as the one described by Pesidiris [62] can also be implemented more easily with more accurate pulsating flow simulations. Furthermore, better predictions in the turbine outlet line can lead to a better design of the aftertreatment and silencer, and, consequently, to an improvement in the consumption and pollutant emissions.

A lot of improvements have been seen in this area during the last decade, as shown in the works by Romagnoli et al. [63] or Payri et al. [64]. Furthermore, some advances developed in the classic unidimensional approach can be found in the works of Chen et al. [24], Costall et al. [35] or De Bellis et al. [65]. Other aspects of the turbocharger behaviour can be taken into account in these simulations. Such is the case of the heat transfer effects, that can be determined experimentally with a simple lumped model as in [66] or [67] later in [27] or [68], and more recently in [69] . The simplifications that are used in simple turbocharger heat transfer models that are coupled with one-dimensional gas dynamics codes have been validated both experimentally and with computational fluid dynamics (CFD) simulations, as in [70] or in [71], where a Conjugate Heat Transfer (CHT) simulation is used to characterise the heat transfer in the turbine. The results of including a heat transfer model to determine internal heat flows has been investigated by several authors, such as in the works by Olmeda et al. [72] or Serrano et al. [73], and with more focus in the compressor performance in [74]. A review of the effect of heat transfer in turbochargers can be found in [75]. These models can even be used to obtain information that is difficult to measure in an experiment, such as the difference heat flow paths, as in the work by Aghaali et al. [76]. Friction losses models as the one developed by Serrano et al. [32] or Marelli et al. [77] have been also coupled with one-dimensional gas dynamics codes so the mechanical efficiency can be computed in time marching simulations, as shown in [78]. Several works about radial turbine modelling can be found in the literature, where the properties of the internal flow of the turbine under different boundary conditions are presented and discussed, as can be seen in the works by Galindo et al. [79, 80] and Hakeem et al. [81].

Especially focused in the unsteady performance at high frequencies are the works by Hu [82] and King [83]. In addition, the one-dimensional unsteady flow model presented by Feneley et al [84] takes into consideration the circumferential feeding of the rotor channels that are also treated as one-dimensional and unsteady and solved by means of the method of characteristics. Also, the experimental characterisation of turbochargers is increasing in complexity in parallel to the models that use this information. This can be seen in the works by Rajoo et al. [85] or Hohenberg et al. [86] for pulsating flow in turbines, Serrano et al. [87] for off-design performance of turbines or Torregrosa et al. [88], Leufvén and Eriksson [89] and Galindo et al. [90] for the performance, including noise emissions and off-design working conditions, of radial compressors.

In the current chapter of the book, a quasi-bidimensional approximation is proposed with the intention of improving the current performance at high frequencies of the equivalent one-dimensional duct approach with the minimal penalty in computational costs.

The objectives of this chapter are to present a method to get pressure data in a radial turbine and to fit and validate a simple quasi-two-dimensional turbine model, derived from [91], with that experimental data. This model is able to take into account the circumferential non-uniformity of the flow at the volute outlet section, which is an important feature to capture the high-frequency performance of the turbine, as shown by Ding et al. [92] [93].

2.2 Model description and method of calculation

In this section, the main aspects of the model will be described and the specific method of calculation will be presented. It is important to mention that the model presented here is an evolution of a pure 1D model, described in [24] and sketched in Figure 2.1. Section 0 represents the domain inlet section, section 1 is turbine inlet, section 2 is the equivalent stator nozzle, section 3 is the rotor inlet, section 4 is the rotor outlet, section 5 is the turbine outlet and section 6 is the domain outlet.

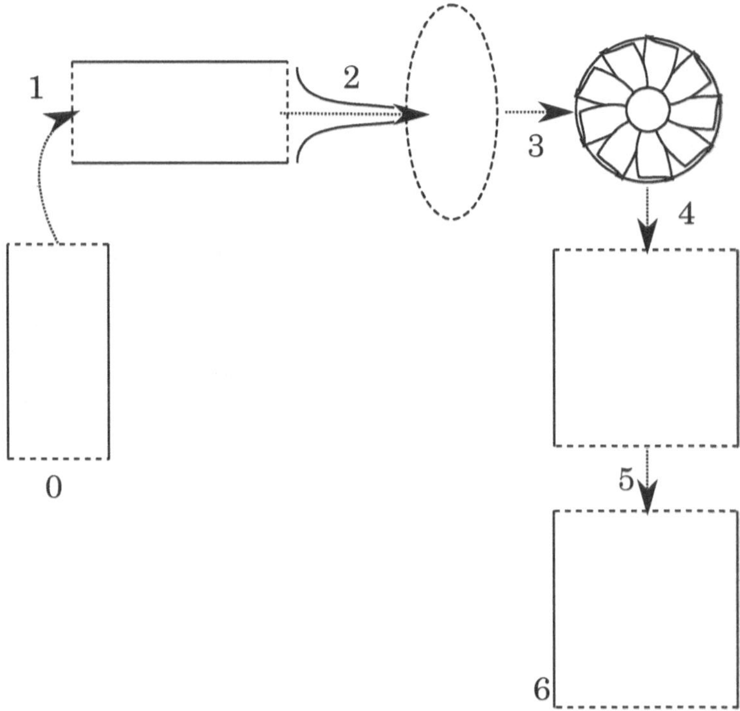

Figure 2.1: Previous 1D Model. The main sections are labelled with numbers.

The model is divided in several parts: one-dimensional flow ducts (turbine inlet and outlet) and a quasi-bidimensional duct (volute). The stator and the rotor are modelled as equivalent nozzles and are computed using variable equivalent areas and a virtual plenum between them. This plenum has a volume equal to that of the physical stator and rotor combined, to take into account the mass and energy accumulation effects. The model schematic is shown in Figure 2.2, where the same numbering for the different sections is used as in the case of the pure 1D model.

The simulation results that are presented in this study were obtained using Open-WAM [17], the 1D simulation integrated tool developed in CMT–Motores Térmicos. Euler classical governing equations for fluid dynamics are discretised across the set of 0D and 1D elements of the computational domain. The set of equations that are discretised and solved are shown in Equation 2.1 and Equation 2.2.

$$\mathbf{w} = \begin{pmatrix} \rho \\ \rho \cdot u \\ \rho \cdot e_t \end{pmatrix} = \begin{pmatrix} \rho \\ \rho \cdot u \\ \rho \cdot c_v \cdot T + \rho \cdot u^2/2 \end{pmatrix} \tag{2.1}$$

$$\frac{\partial \mathbf{w}}{\partial t} = \frac{\partial}{\partial x} \begin{bmatrix} \rho \cdot u \\ \rho \cdot u^2 + p \\ \rho \cdot u \cdot (c_p \cdot T + u^2/2) \end{bmatrix} \tag{2.2}$$

where $\mathbf{w}$ is the state vector, ρ is the density, u is the flow speed, e_t is the specific total internal energy, c_v is the isochoric specific heat capacity, T is the fluid temperature, t is the time, p is the gas pressure and c_p is the isobaric specific heat capacity. An ideal gas equation of state is also used, as in Equation 2.3:

$$p = \rho \cdot R \cdot T \tag{2.3}$$

where R is the gas constant. The older two-step Lax-Wendroff scheme in finite differences [94] was replaced by a Finite Volume Method, leading to a better result in terms of mass, momentum and energy conservation. After discretising the equations by a Finite Volume Method approach, Equation 2.4 is obtained:

$$\frac{\mathrm{d}\bar{\mathbf{w}}_i}{\mathrm{d}t} = \frac{(A_{i-1,i} \cdot \mathbf{F}_{i-1,i} - A_{i,i+1} \cdot \mathbf{F}_{i,i+1} + \mathbf{C}_i)}{V_i} \tag{2.4}$$

where $\bar{\mathbf{w}}_i$ is the mean value of the state vector in the cell i, $A_{i-i,i}$ is the boundary surface between cells $i-1$ and i, $\mathbf{F}_{i-1,i}$ is the flux vector between cells $i-1$ and i and $\mathbf{C}_i$ is the source terms vector affecting cell i. Finally, V_i is the volume of the cell. The source term is computed as in Equation 2.5 when no mass or heat sources are present:

$$\mathbf{C}_i = \begin{bmatrix} 0 \\ p_i \cdot (A_{i-1,i} - A_{i,i+1}) \\ 0 \end{bmatrix} \tag{2.5}$$

A Godunov-derived scheme reconstruction ([95]) was implemented, using a second order Monotone Upstream-centered Scheme for Conservation Laws (MUSCL) [96] for improving the spatial accuracy during the integration. The linear extrapolation of the state vector to the cell boundaries is limited using a monotonised central slope limiter by van Leer, and the fluxes are computed using the Harten-Lax-van Leer-Contact approximate Riemann solver [21].

The time-step is obtained at each iteration following the Courant-Friedrichs-Lewy (CFL) condition [97], as in Equation 2.6:

$$\Delta t = \nu \cdot \left. \frac{\Delta x}{\lambda} \right|_{\min} = \nu \cdot \left. \frac{\Delta x}{|u| + |a|} \right|_{\min} \tag{2.6}$$

where ν is the Courant number, Δx is the cell size, λ is an eigenvalue of the system, a is the speed of sound and the subscript min represents the minimum value. The Courant number used during the simulations was 1.0. After computing all the terms, Equation 2.4 is solved each time-step by means of Heun's method.

In the case of the volute, the same methodology has been applied but taking into account that, although the volute is simulated as a one dimensional tapered duct, source terms will be considered in the 1D conservation equations due to the flow leaving through the lateral window. The addition of these source terms, computed using the stator results, allows the model to capture the non-uniformity of the flow in the volute during the pulse transmission. A more realistic volute geometry is used in this case: instead of a simple straight duct with half of the actual turbine length, the real section distribution and length are used.

The source term $\dot{m}_{\mathrm{cell}}$ for the mass conservation equation for a volute cell is shown in Equation 2.7:

$$\dot{m}_{\mathrm{cell}} = \dot{m}_{\mathrm{st}} \tag{2.7}$$

where $\dot{m}_{\mathrm{st}}$ is the mass flow that passes through one of the stator channels, using as total inlet conditions the total pressure and temperature of the volute cell, and is computed as described in the following paragraphs. The source term $\dot{H}_{\mathrm{cell}}$ for the energy conservation equation for a volute cell is shown in Equation 2.8:

$$\dot{H}_{\mathrm{cell}} = \dot{m}_{\mathrm{st}} \cdot c_{\mathrm{p}} \cdot T_{\mathrm{cell,t}} \tag{2.8}$$

where c_{p} is the isobaric specific heat capacity and $T_{\mathrm{cell,t}}$ is the total temperature of the cell. Finally, the source term $\dot{q}_{\mathrm{cell}}$ for the momentum conservation equation is computed as in Equation 2.9:

$$\dot{q}_{\mathrm{cell}} = \dot{m}_{\mathrm{st}} \cdot v_{\mathrm{cell}} \tag{2.9}$$

where v_{cell} is the axial flow velocity in the cell.

Regarding the prediction of the flow variation around the volute, the work by Chiong et al. [39] proposes a different solution based on a non-adiabatic pressure loss boundary condition. Also in [91], a description of the quasi-bidimensional model of the volute can be found, however, the stator and rotor are not modelled as in this book, as it will be next explained.

The stator is solved as several nozzles connected to the volute, each to a volute cell, and to the stator outlet discharge zone using several equivalent effective areas. This way, the non-uniformity of the flow properties in the volute outlet section can be taken into account. On the other hand, the rotor is modelled using one equivalent nozzle with a determined equivalent effective area. Once the effective areas are known, the computation of the real flow through the stator and through the rotor is easily obtained by considering adiabatic behaviour in the nozzles. The procedure to obtain these effective areas is described in the calibration section.

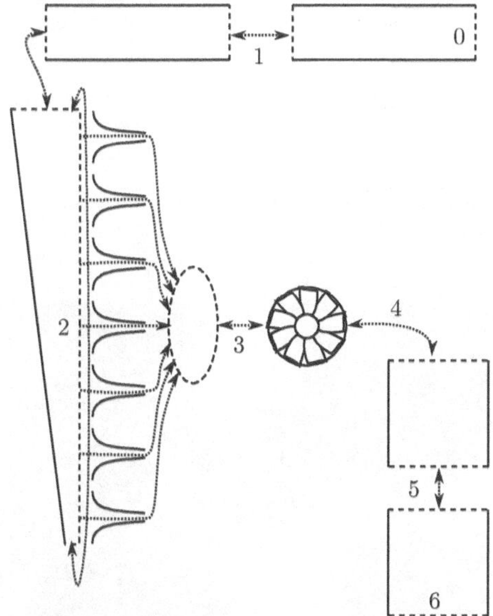

Figure 2.2: Q2D Model schematic. The main sections are numbered.

In order to simplify the validation process, the flow evolution is supposed to be adiabatic, so quasi-adiabatic experimental data is needed. Some basic geometrical data will also be necessary. The following geometrical data is used by the proposed model:

- Ducts and volute lengths and section distribution.

- Stator blades sizes and positions.

- Stator inlet and outlet radii and height.

- Rotor outlet tip and nut radii.

- Rotor outlet metal angle.

The volute itself has been used as a mould in which silicone has been injected in order to get its geometry. After the silicone had cured, it was extracted and measured. The mould can be seen in Figure 2.3. Not long time after, a 3D scanner was used to gather more accurate information of the geometry. Then, the data obtained was later used for CFD studies and applied to develop a new turbine efficiency extrapolation model [98]. However, the work carried out in this book uses the more mature extrapolation model instead [30].

After that, several photographs were taken and a vernier calliper used to measure the rotor diameters and the rotor outlet angle, as portrayed in Figure 2.4. The inlet and outlet wheel diameter are 41 mm and 38 mm respectively.

Finally, more photographs where taken to measure the position of the stator vanes, their lengths and angles for different VGT displacements. The angle of each vane was individually measured and their mean value was used for each VGT displacement.

Figure 2.3: A silicone mould of the volute.

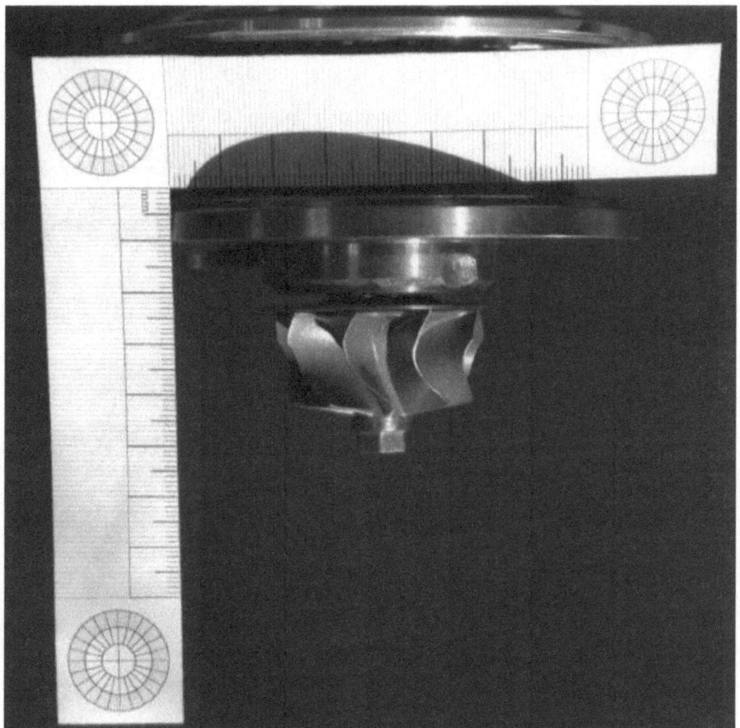

Figure 2.4: The turbine rotor wheel.

2.3 Calibration method

In this subsection, the model fitting will be described in detail along with the equations that are needed. To succeed in the fitting process, turbine mass flow rate, inlet total conditions and stator static pressure during steady tests must be measured.

Firstly, to capture the unsteady phenomena in any engine working condition, the model needs information of the reduced mass flow and the adiabatic efficiency in the complete turbine operation range, so the model map needs to be first *adiabatized* and then extrapolated. For the adiabatization process, the residual heat losses need to be computed and extracted [27, 28, 29]. On the other hand, the extrapolation procedure described in detail in [30] is based on modelling the turbine as a single equivalent nozzle.

27

This method, as it will be next synthesised, is able to extrapolate the radial turbine performance in terms of VGT position, rotational speed and blade speed ratio, using thirteen calibration coefficients; seven for the reduced mass flow and six for the efficiency.

In the case of the reduced mass flow extrapolation procedure, the starting point is the use of continuity applied to stator, rotor and equivalent nozzle as shown in Equation 2.10. Regarding the equivalent nozzle area, using the velocity definition and solving the mass flow Equation 2.11 is obtained, and, assuming certain hypothesis (as shown in the work by Serrano et al. [30]), a new equation of the throat area of the equivalent nozzle is deduced 2.12:

$$\dot{m} = A_{2'}\rho_{2'}v_{2'} = A_4\rho_4 w_4 = A_{\mathrm{Neq}}\rho_4 v_{\mathrm{Neq}} \tag{2.10}$$

$$A_{\mathrm{Neq}} = A_4 \sqrt{\frac{1 + \left(\frac{u_4}{v_{\mathrm{Neq}}}\right)^2 - \left(\frac{u_3}{v_{\mathrm{Neq}}}\right)^2 + \left(\frac{w_3}{v_{\mathrm{Neq}}}\right)^2}{\left(\frac{A_4}{A_{2'}}\right)^2 \left(\frac{\rho_4}{\rho_{2'}}\right)^2 + 1}} \tag{2.11}$$

$$A_{\mathrm{Neq}} = \frac{a \cdot A_4 \cdot \sqrt{1 + \dfrac{\sigma^2 \cdot \left[\left(\frac{D_4}{D_3}\right)^2 - 1\right] + b}{\overline{\overline{\eta}}_{\mathrm{t,s}}}}}{\sqrt{1 + \left(c \cdot \dfrac{A_4}{A_{2'}}\right)^2 \cdot \dfrac{\left(\frac{1}{\pi_{2',4}}\right)^2}{\left\{1 - \eta_{\mathrm{t,s}} \cdot \left[1 - \left(\frac{1}{\pi_{2',4}}\right)^{\frac{\gamma-1}{\gamma}}\right]\right\}^2}}} \tag{2.12}$$

The A_{Neq} is computed minimizing the difference with the equivalent effective area from the turbine map with the method of least squares of Levenberg-Marquart. The coefficients a, b and c are fitted according to the following:

- $a = cte$

- $b = b_0 + b_1 \cdot VGT$

- $c = c_0 + c_1 \cdot VGT$

where a is a constant value subjected to constrains and not dependent on the VGT.

It is important to point out that Equation 2.12 depends only on easy measurable geometry parameters of the turbine and on the information available in a standard map. Once the equivalent nozzle area A_{Neq} is known, the reduced mass flow can be calculated considering flow through an orifice with isentropic expansion:

$$
\dot{m}_{\mathrm{red}} = A_{\mathrm{Neq}} \sqrt{\frac{\gamma}{\mathfrak{R}}} \left(\frac{1}{\pi_{1\mathrm{t},4\mathrm{s}}} \right)^{\frac{1}{\gamma}} \cdot \sqrt{\frac{2}{\gamma - 1} \left[1 - \left(\frac{1}{\pi_{1\mathrm{t},4\mathrm{s}}} \right)^{\frac{\gamma-1}{\gamma}} \right]}
\tag{2.13}
$$

The model takes into consideration if critical conditions are reached by means of limiting the expansion ratio (as in Equation 2.13), detecting independently the possible choking flow in stator or rotor. On the other hand, to calculate the adiabatic efficiency in Equation 2.14, the Euler equation and the turbine enthalpy drop are used. Through this procedure, Equation 2.15 is reached. However, as proved in Serrano et al. [30], Equation 2.16 can be deduced, which is built with the fitting parameters. The efficiency of the model is, thus, obtained with a similar procedure as the one needed for the effective area.

$$
\eta_{t,s} = \frac{T_{1\mathrm{t}} - T_{4s}}{T_{1\mathrm{t}} - T_{4s,\mathrm{isentropic}}}
\tag{2.14}
$$

$$
\eta_{t,s} = -2 \left(\frac{D_4}{D_3} \right)^2 \sigma^2 + 2 \frac{A_{\mathrm{Neq}}}{A_1} \left(\tan \alpha_3 + \frac{D_4}{D_3} \tan \beta_4 \right) \cdot \left[\frac{1}{\pi_{1\mathrm{t},4\mathrm{s}}} \right]^{\frac{1}{\gamma}} \sigma
\tag{2.15}
$$

$$
\eta_{t,s} = -d\sigma^2 + e \left(1 - \frac{f}{\sigma^2} \right)^{\frac{1}{\gamma-1}} \cdot \sigma
\tag{2.16}
$$

As a result, Equation 2.12 and Equation 2.16 can be used for the non-linear fitting process of the coefficients and they should be solved iteratively as a system, since the reduced mass flow appears in the efficiency equation and the efficiency equation appears in the reduced mass flow equation. To succeed in the fitting procedure, some boundary conditions for the coefficients must be considered, that way, the definition of each coefficient has a physical explanation. With the fulfilment of this procedure, demonstrated in [30], the radial turbine performance has been extrapolated and an effective area of the equivalent nozzle has been characterised for a wide range of operating points.

Once the extrapolated map is obtained as an output of the model, the equivalent effective area and efficiency are chosen in each time-step computation following two possibilities. If the VGT position falls inside the map, the effective area and efficiency are selected by means of a linear interpolation between the immediately superior and inferior values provided by the model. On the contrary, if the VGT position is outside the map, a weighted averaged method is used between the closest value of the map and the model fitted to the complete map.

It is commonplace to consider a value for the reaction degree of R = 0.5 to model the behaviour of radial turbines designed without guide blades. Nevertheless, that hypothesis cannot be generally accepted when modelling variable geometry radial turbines. For that reason, an additional procedure, detailed in [25], is applied in order to calculate the pressure drop across the turbine stator and rotor and thus, evolve the model up to two nozzles with an intermediate volume between them. In summary, the intermediate pressure can be expressed as:

$$\frac{p_3}{p_1} = \left(1 + \frac{T_{1t}}{T_1}(R-1)\eta_{t,s}\left[1 - \left(\frac{p_4}{p_{1t}}\right)^{\frac{\gamma-1}{\gamma}}\right]\right)^{\frac{\gamma-1}{\gamma}} \tag{2.17}$$

where

$$R = 0.5 \qquad \text{if } \alpha_1 > \alpha_{\text{limit}} \tag{2.18}$$

and

$$R = 1 - \frac{2 \cdot \Re}{\pi^2 D_3 \left(D_4^2 - D_{4,nut}^2\right)} \cdot \frac{\dot{m}_{\text{red}} \cdot \tan \alpha_3}{N_{\text{red}}}$$
$$\cdot \left(\frac{p_4}{p_{1t}}\right)^{-1} \left\{\left(\frac{T_{1t}}{T_1}\right)^{-1} - \eta_{t,s}\left[1 - \left(\frac{p_4}{p_{1t}}\right)^{\frac{\gamma-1}{\gamma}}\right]\right\} \tag{2.19}$$

otherwise.

Note that in the previous equations the only inputs are geometrical parameters, along with the typical variables provided by turbochargers manufacturers. Furthermore, instead of a variable polytropic coefficient, isentropic evolution has been considered in order to simplify the calculations. Once the pressure drop across the stator is calculated, the effective areas of the nozzles equivalent to the stator and the rotor can be obtained from Equation 2.20 and Equation 2.21, where $p_{1t,\text{rel}}$ is the relative condition at the rotor inlet. In addition, to obtain the term $p_{3t,\text{rel}}/p_4$, Equation 2.22 must be computed.

$$A_{\text{eff,st}} = \dot{m} \cdot \frac{\sqrt{\Re\gamma T_{1t}}}{p_{1t}} \cdot \frac{1}{\gamma} \cdot \left(\frac{p_{1t}}{p_3}\right)^{1/\gamma}$$
$$\cdot \left\{\sqrt{\frac{2}{\gamma-1}\left[1 - \left(\frac{p_3}{p_{1t}}\right)^{\frac{\gamma-1}{\gamma}}\right]}\right\}^{-1} \tag{2.20}$$

$$A_{\text{eff,rot}} = \dot{m} \cdot \frac{\sqrt{\Re \gamma T_{3t}}}{p_{3t}} \cdot \frac{1}{\gamma} \cdot \left(\frac{p_{3t,\text{rel}}}{p_4}\right)^{1/\gamma}$$
$$\cdot \left\{ \sqrt{\frac{2}{\gamma-1}\left[1 - \left(\frac{p_4}{p_{3t,\text{rel}}}\right)^{\frac{\gamma-1}{\gamma}}\right]} \right\}^{-1} \tag{2.21}$$

$$\frac{p_{3t,\text{rel}}}{p_4} = \frac{p_3}{p_1} \cdot \frac{p_{1t}}{p_4} \cdot \left(\frac{T_1}{T_{1t}}\right)^{\frac{\gamma}{\gamma-1}} \cdot \left[(1 + \left(\frac{T_{1t}}{T_1} - 1\right) \cdot \frac{T_1}{T_3}\right]^{\frac{\gamma}{\gamma-1}} \tag{2.22}$$

Finally, with the rotor and stator equivalent areas, the mass flow can be obtained taking into account that the total temperature is conserved when considering isentropic nozzle flow. The expansion ratio of each volute cell is used for the computation in each time step. Therefore, each lateral nozzle has its own expansion ratio and effective area to reproduce the volute mass flow in the rotor entry. In addition, each volute cell contributes to the rotor total effective area, that will be obtained by averaging the ones obtained in all the volute cells. Regarding the turbine extracted power, it is computed as a sum of the power of each of the volute cells. Again, the explanation of the hypotheses and the demonstrations of the previous equations can be found in [25].

2.4 Experimental setup

The experiments have been done in a turbocharger gas stand with the following arrangement:

- A screw compressor with a maximum flow rate of $0.2 \, \text{kg s}^{-1}$ and a maximum discharging absolute pressure of $0.4 \, \text{MPa}$ provides the mass flow to the turbine. This mass flow rate is controlled by the screw compressor speed and an electronic discharge valve placed after the screw compressor. The discharge valve is used when a very small mass flow rate and turbine inlet pressure is required.

- The temperature of the flow that goes to the turbine can be:
    - Cooled by a water-air heat exchanger.
    - Heated by five electrical heaters with an installed power of $40 \, \text{kW}$, rising the turbine inlet temperature up to $720 \, \text{K}$ at maximum flow rate.

- The flow that enters the turbocharger compressor passes through a filter and a water-air heat exchanger in order to regulate the compressor inlet temperature.

- The turbocharger compressor flow is controlled by means of a backpressure valve.

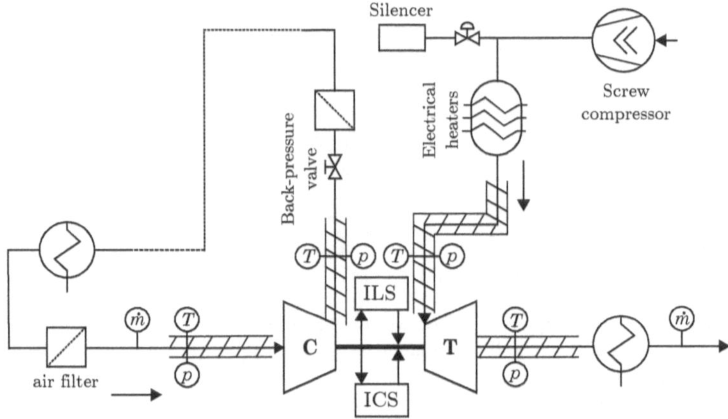

Figure 2.5: Gas stand schematic.

- The compressor side of the gas stand can be operated in closed-loop, so the flow can be pressurised. This leads to wider operating ranges when using the turbocharger compressor as a braking system while measuring turbine maps.

- A rotary valve can be placed upstream of the turbine or downstream of the compressor to generate pulses similar to that found in a reciprocating engine.

- An independent lubrication system is used for the turbocharger. The oil pressure and temperature can be adjusted for each experimental test.

- A hot film flow meter is placed at the turbine side gas stand outlet, after a water-air heat exchanger. This mass flow meter has an expanded uncertainty of 2 % of the measured value in the worst case for the measured values during the experimental campaign.

- Type K thermocouples, with an expanded uncertainty of 1.5 K per thermocouple, are used to measure air temperatures at the turbine inlet and outlet in all the experimental tests.

- Piezoresistive transmitters are used to measure the static air pressure at the turbine inlet and outlet.

- Three piezoresistive transducers are used to measure instantaneous static pressure at the turbine inlet and outlet.

- An induced currents transducer is placed at the compressor side to monitor the turbocharger speed.

- A displacement transducer is used to measure the VGT position. The maximum opening corresponds to a VGT position of 100 % and the minimum to 0 %. Using the straight-line method all the intermediate positions are known.

- All the sensors are installed according to Society of Automotive Engineers (SAE) J1723 and J1826 standards [99, 100].

In Figure 5.1 the gas stand schematic used during the tests is presented. The internal cooling system (ICS) and the internal lubrication system (ILS) are also provided with pressure and temperature sensors, along with mass flow sensors, but they have been omitted for clarity's sake. In Table 2.1 the main gas stand measurement equipment is summarised with the corresponding uncertainty and the operating range.

Variable	Sensor type	Range	Typical expanded uncertainty
Gas mass flow	Thermal flowmeter	$45\,\mathrm{kg\,h^{-1}}$ to $1230\,\mathrm{kg\,h^{-1}}$	$< 1\,\%$ of reading output
Turbocharger speed	Inductive sensors	$0\,\mathrm{krpm}$ to $300\,\mathrm{krpm}$	$< 500\,\mathrm{rpm}$
Gas temperature	Type K thermocouple	$273\,\mathrm{K}$ to $1500\,\mathrm{K}$	$1.5\,\mathrm{K}$
Average gas pressure	Piezoresistive transmitter	$0\,\mathrm{MPa}$ to $0.5\,\mathrm{MPa}$, absolute	$12.5\,\mathrm{hPa}$
Turbine vanes position	Inductive displacement sensor	$0\,\%$ to $100\,\%$	$< 1\,\%$
Oil temperature	Platinum RTD	$173\,\mathrm{K}$ to $723\,\mathrm{K}$	$< 0.5\,\mathrm{K}$
Oil pressure	Piezoresistive transmitter	$0\,\mathrm{MPa}$ to $0.5\,\mathrm{MPa}$	$12.5\,\mathrm{hPa}$
Oil mass flow	Coriolis flow meter	$0\,\mathrm{g\,s^{-1}}$ to $50\,\mathrm{g\,s^{-1}}$	$2\,\%$ of reading output

Table 2.1: Gas stand measurement equipment.

The heat flow has been maintained as low as possible during the experimental campaign, in what is commonly called quasi-adiabatic conditions. To ensure it, the turbine and lubrication inlet temperatures have been set to be close to the compressor outlet temperature, with a maximum allowed discrepancy of $\pm 6\,\mathrm{K}$ and a mean value of $\pm 2.5\,\mathrm{K}$. Also, as the compressor outlet temperature is relatively low and the ducts were thermally insulated, the heat flow to the environment is bounded to a small amount of around 2 % of the measured turbine power output, as can be seen in Figure 2.6. This heat flow to the environment has been estimated using the environment temperature T_{env}, the maximum surface temperature of the turbine T_{surf} and the exposed surface of the ducts and turbine

A_{surf}, and assuming a convective heat transfer coefficient h_{conv} of $10\,\text{W}\,\text{m}^{-2}\,\text{K}^{-1}$:

$$\dot{Q}_{\text{env}} = A_{\text{surf}} \cdot h_{\text{conv}} \cdot (T_{\text{surf}} - T_{\text{env}}) \tag{2.23}$$

The turbine power output is estimated using the total enthalpy flow as follows:

$$\dot{W} = \dot{m} \cdot c_{\text{p}} \cdot (T_{6\text{t}} - T_{0\text{t}}) - \dot{Q}_{\text{int}} \tag{2.24}$$

where the term $\dot{Q}_{\text{int}}$ corresponds to the internal heat losses from the gas to the turbine shroud, and is calculated as described in [67].

In the next figure is shown the expected error in percentage in the calculation of the turbine power output due to the external heat losses.

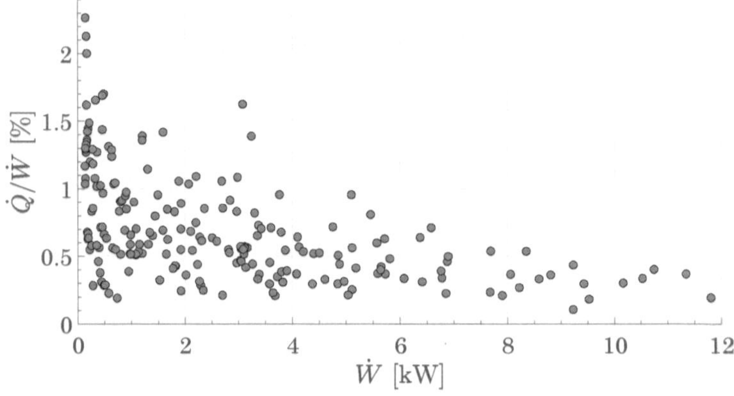

Figure 2.6: Expected error in turbine power output due to external heat flow effects.

2.4.1 Experimental campaign details

The experimental campaign consisted of two clearly differentiated parts, the quasi-adiabatic test and the pulsating test. In the quasi-adiabatic test the objective was to obtain a complete turbine map and all the necessary reliable data for characterising the turbine and perform the validation of the model under pulsating conditions. As it was commented before, the pulsating flow is generated by a rotary valve placed upstream of the turbine. A disc with several windows is placed at the turbine line in a way that it opens and closes the air flow path as it rotates. In order to adjust the pulse amplitude, a bypass valve was used in parallel with the rotating valve. On the other hand, the frequency of the pulse was controlled varying the frequency of the electrical engine power supply from 0 Hz to 50 Hz. The law of aperture of the disc was designed to approximate the behaviour of that found in the exhaust manifold of a four cylinder, four strokes reciprocating engine. Taking into account that the disc had three windows, it was possible to simulate a up to 4500 rpm. Several engine configurations were simulated in the gas stand:

- 2000 rpm, 25 % of maximum BMEP

- 2000 rpm, 100 % of maximum BMEP

- 3000 rpm, 50 % of maximum BMEP

- 3000 rpm, 75 % of maximum BMEP

- 3000 rpm, 100 % of maximum BMEP

- 3500 rpm, 100 % of maximum BMEP

When designing the experimental campaign, the aim was to obtain a representative set of data for a complete range of operating conditions. In Table 2.2, the test matrix used during the experiments is shown:

Engine speed [rpm]	Load [%]	Pulse frequency [Hz]	VGT position [%]	Turbo speed [krpm]
2000	25	66.67	56	58
2000	100	66.67	47	110
3000	50	100	80	99
3000	75	100	60	122
3000	100	100	50	123
3500	100	116.67	55	124

Table 2.2: Pulsating flow test matrix for single entry turbine experiments.

2.4.2 Pressure decomposition

In order to obtain the instantaneous pressure results, a matrix of three pressure transducers was fixed in both inlet and outlet branch, denoted by p in Figure 2.7.

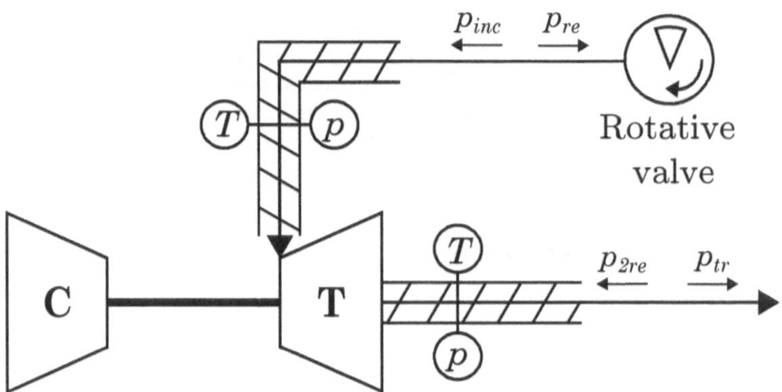

Figure 2.7: Pulsating test schematic. Incident and 2nd reflection waves are imposed.

Along with the instantaneous pressure measurements, piezoresistive sensors have been used for measuring the static mean pressure in the inlet and outlet branch, taking into account that piezoelectric transducers don't measure the very low frequency components of the pressure. On the other hand, because of the accuracy and time-resolution requirements, piezoelectric sensors have been used for the instantaneous pressure measurements.

In order to study the pressure waves and obtain the incident and reflected components, a beamforming technique has been used [101]. This method uses spatial filtering techniques to a set of instantaneous pressure measurements in order to estimate the velocity fluctuations through the decomposition of the flow into its forward and backwards components, assuming that the flow speed is equal to the linear superposition of the forward and backward flow velocities. To perform correctly this technique, as well as to avoid possible aliasing phenomena, the flow needs to be fully developed. To achieve that, it is necessary to fix a minimum distance of approximately 10 pipe diameters of unobstructed straight length in both directions, upstream and downstream of the sensor matrix array. Furthermore, the three pressure sensors that compose a beamforming array for each of the three turbine branches require a particular physical separation for a correct decomposition calculation. As it can be observed in the sketch from Figure 2.8, this distance between sensors was set to 5 cm, which has proven to be a good compromise between the measurement precision and the linear propagation of the waves between the sensors.

To capture and process the signal generated by the sensors, charge amplifiers were used with a calibrated data acquisition unit operating at a sampling frequency of 100 kHz. Every sensor was previously calibrated for adjusting the sensitivity.

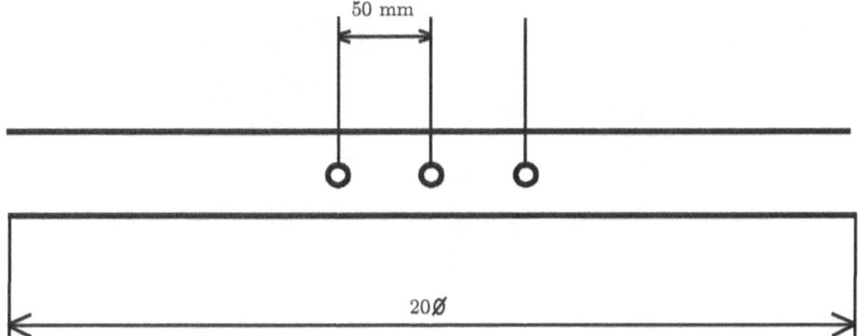

Figure 2.8: Piezoelectric instantaneous pressure sensors sketch.

According to the method used, the level of entropy A_A, that can be defined as :

$$A_A = \sqrt{\frac{T}{Tref} \cdot \left(\frac{p_{ref}}{p}\right)^{\frac{\gamma-1}{2\cdot\gamma}}}$$

(2.25)

(where T_{ref} is a reference temperature and p_{ref} is a reference pressure) is considered constant during each pulsating test, which implies an isentropic relationship between the instantaneous temperature and pressure at the turbine inlet and outlet.

On the other hand, the flow speed associated to the forward and backwards waves, i.e., p_{right} and p_{left} are defined as:

$$u_{right} = \frac{2 \cdot a_{ref}}{\gamma - 1} \cdot \left[\left(\frac{p_{right}}{p_{ref}}\right)^{\frac{\gamma-1}{2\cdot\gamma}} - 1\right] \cdot A_A$$

(2.26)

$$u_{left} = -\frac{2 \cdot a_{ref}}{\gamma - 1} \cdot \left[\left(\frac{p_{left}}{p_{ref}}\right)^{\frac{\gamma-1}{2\cdot\gamma}} - 1\right] \cdot A_A$$

(2.27)

being a_{ref} the speed of sound at the reference temperature T_{ref}. In these conditions, the instantaneous mass flow rate can be approximated with the next expression:

$$\begin{aligned}
\dot{m} &= \rho \cdot A \cdot u = \rho \cdot A \cdot (u_{right} + u_{left}) \\
&= \frac{p}{R \cdot T} \cdot A \cdot \frac{2 \cdot a_{ref}}{\gamma - 1} \cdot \left[\left(\frac{p_{right}}{p_{ref}}\right)^{\frac{\gamma-1}{2\cdot\gamma}} - \left(\frac{p_{left}}{p_{ref}}\right)^{\frac{\gamma-1}{2\cdot\gamma}}\right] \cdot A_A
\end{aligned}$$

(2.28)

The pressure and temperature sensors were placed at the same radial position but different azimuths, with differences in the range of a few kPa and less than 1 K respectively. As the measurement uncertainty was of the same order, their mean value was used for the calibration and validation phase.

2.5 Results

The results for the turbine map can be seen in Figure 2.9 and Figure 2.10. In Figure 2.9, the turbine reduced mass flow rate $\dot{m}^*$ is plotted against the turbine total to static expansion ratio $\pi_{0t,6}$, the markers represent the experimental points, whereas the lines represent the model results. It is worth noting that only five points per speed line were used to calibrate the model. The model has been fitted using the procedure described in section 'Calibration method' using a random subset of one half of the measured points. The map efficiency from experiments and model are confronted in Figure 2.10. The turbine reduced mass flow rate is defined as:

$$\dot{m}^* = \dot{m} \cdot \frac{\sqrt{T_{0t}}}{p_{0t}} \tag{2.29}$$

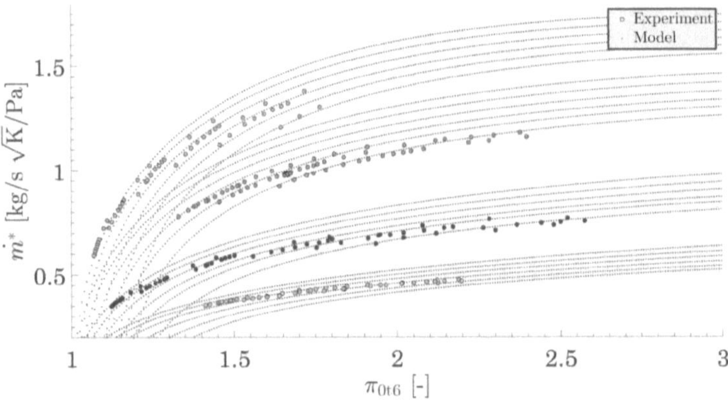

Figure 2.9: Turbine map - measured vs. model.

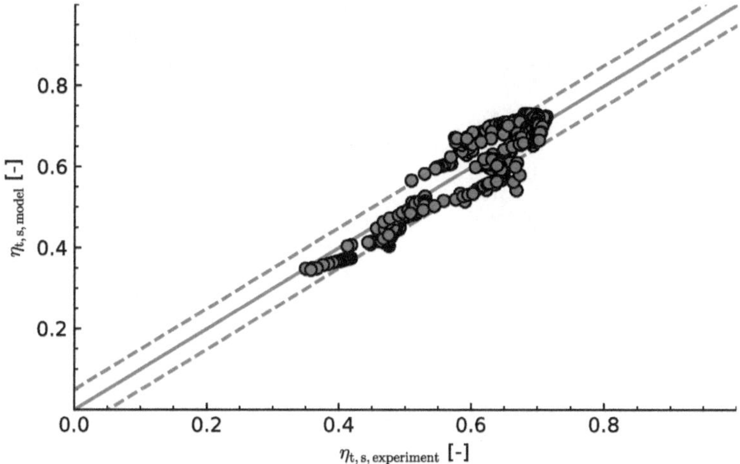

Figure 2.10: Turbine efficiency - measured vs. model.

The instantaneous simulations were carried out imposing the incident wave in the turbine domain inlet (station 0) and the 2nd reflection wave in the turbine domain outlet (station 6) i.e., the locations where the sensors are placed. In order to obtain the incident, reflected, transmitted and 2nd reflection from the composed pressure, beamforming techniques have been used [101].

An example of the results can be seen from Figure 2.11 to Figure 2.14, where the power spectral density of the static pressure is plotted at two different points; the turbine domain inlet (inlet pipe, position 0), and the turbine domain outlet (outlet pipe, position 6) for four different working conditions. The experimental results are plotted against the values provided by the 1D classic volute model and the quasi-bidimensional model. The power spectral density is computed using Welch's average periodogram method [102] using a Hanning window [103] with a 50 % overlap. The two models present similar behaviour in the time domain so, in order to study the differences in the performance between them, the attention has been focused in the frequency domain results. Results are presented for excitations of 66 Hz and 100 Hz, corresponding to 1500 and 3000 rpm.

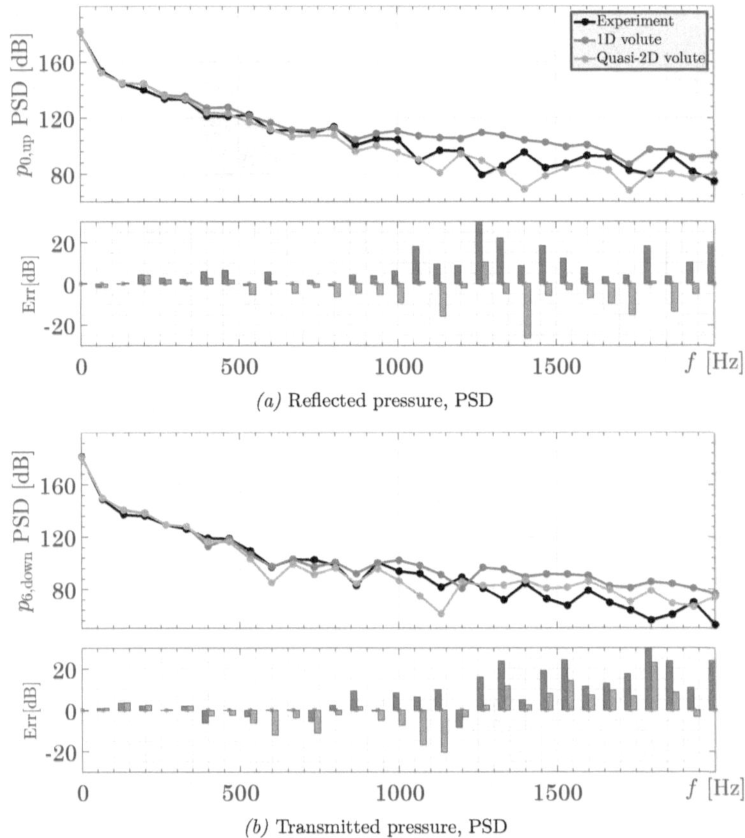

(a) Reflected pressure, PSD

(b) Transmitted pressure, PSD

Figure 2.11: VGT at 56 %, 58 krpm, 66.7 Hz. Reflected and transmitted components.

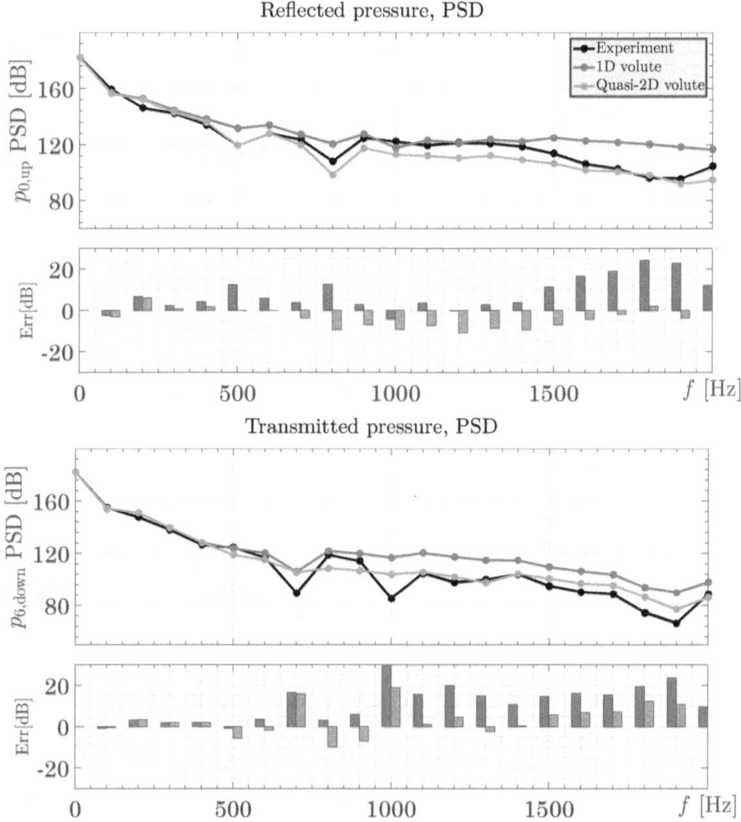

Figure 2.12: VGT at 80 %, 99 krpm, 100 Hz. Reflected and transmitted components.

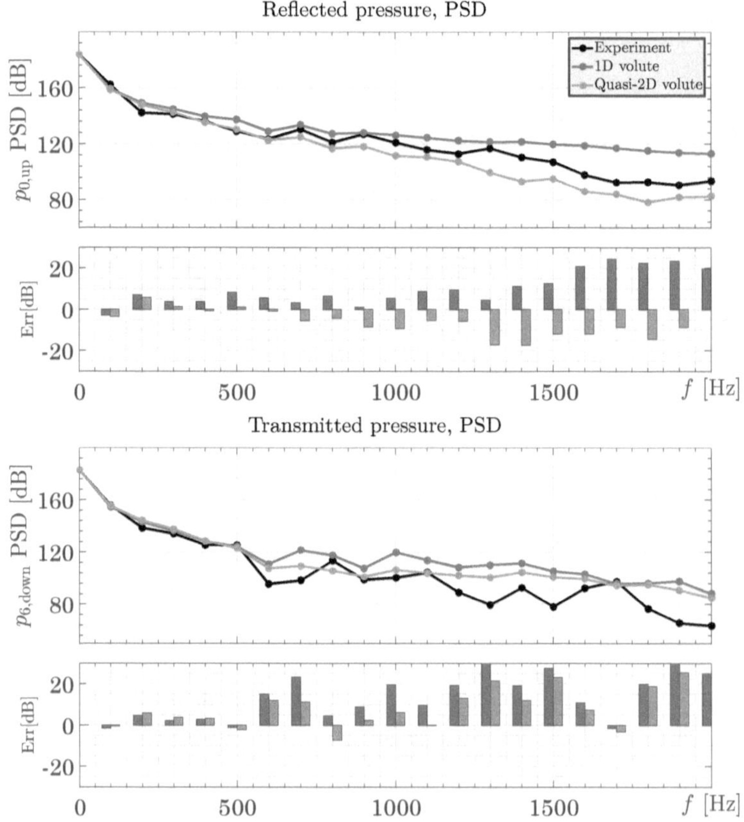

Figure 2.13: First turbocharger, VGT at 60 %, 122 krpm, 100 Hz. Reflected and transmitted components.

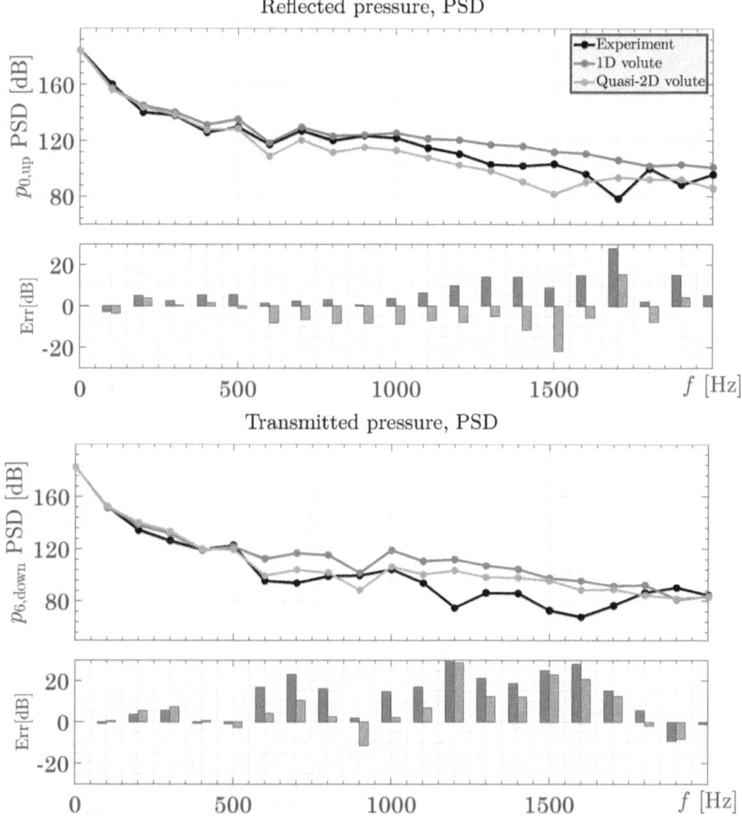

Figure 2.14: First turbocharger, VGT at 50 %, 123 krpm, 100 Hz. Reflected and transmitted components.

Although both models present some discrepancies with the experimental data, the quasi-bidimensional volute shows better fidelity with the experimental results in the range from 1000 Hz to 2000 Hz mainly due to the "averaging" effect as the flow exits through its lateral window. In particular, it can be observed that the 1D volute model tends to overestimate the values in the high frequency range.

From the results it can be inferred that the improvement of the Q2D model is slightly more evident where the pulses have a higher fundamental frequency. This can be explained by the fact that in the higher engine speed cases (higher fundamental frequencies) the wave length reaches the volute length order of magnitude at lower order and higher amplitude harmonics than in the low engine speed cases. Nevertheless, no significant trends were found in terms of VGT position.

On the other hand, for the instantaneous mass flow results, the performance of both 1D and quasi-2D models is similar in time domain as it can be observed in Figure 2.15. This is coherent with the pressure results, as pressure evolution in time domain is similar for both models (see Figure 2.16), and differences only appear in middle and high frequencies as it has been previously indicated.

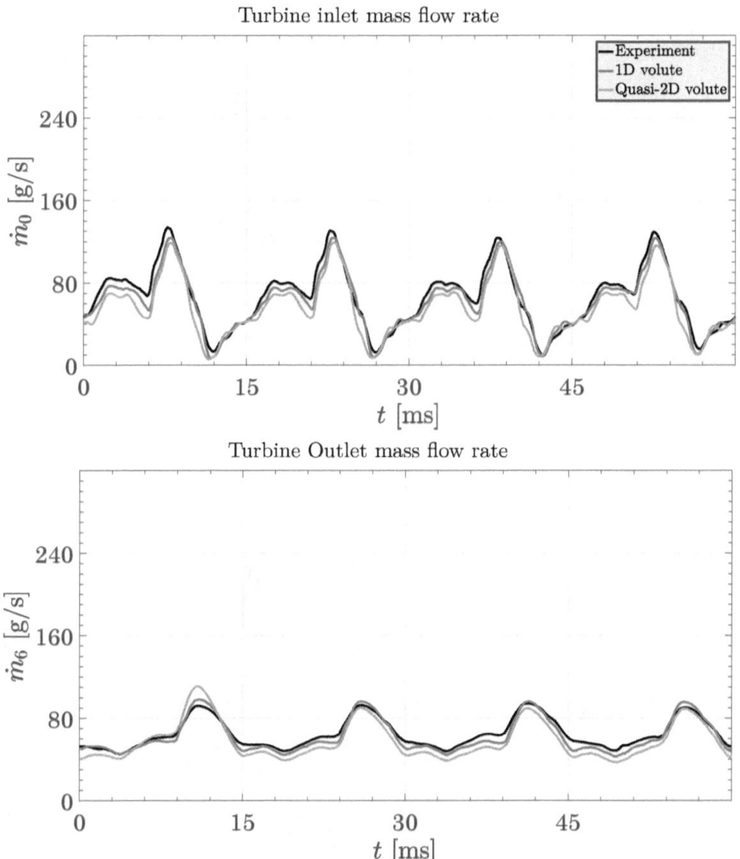

Figure 2.15: First turbocharger, VGT at 50 %, 123 krpm, 100 Hz. Inlet and outlet mass flow rate.

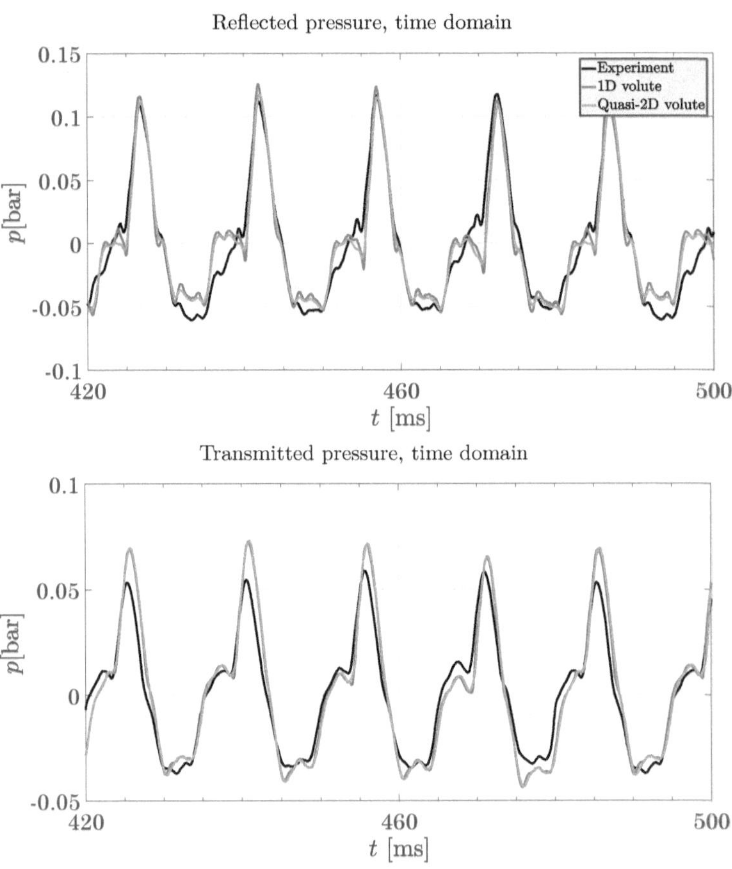

Figure 2.16: First turbocharger, VGT at 50 %, 123 krpm, 100 Hz. Reflected and transmitted in time domain.

Regarding the turbine efficiency comparison, experimental instantaneous efficiency is not available and thus, only the simulation results for both modelling approaches are presented for both instantaneous (Figure 2.17), and pulse-averaged (Figure 2.18).

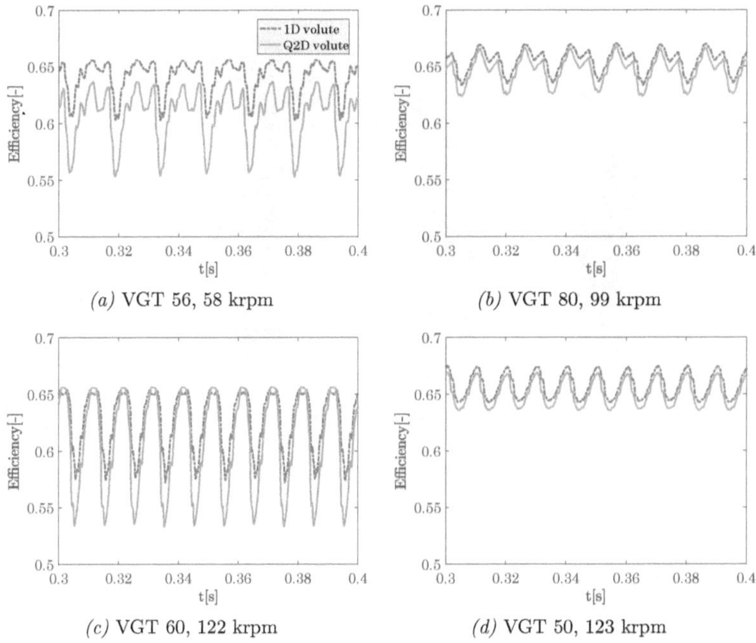

(a) VGT 56, 58 krpm

(b) VGT 80, 99 krpm

(c) VGT 60, 122 krpm

(d) VGT 50, 123 krpm

Figure 2.17: Instantaneous Efficiency.

From the results of instantaneous efficiency it can be observed that there are some noticeable difference between the models, especially in the case of 2.17(a), where the minimum value reached by the 1D simulation is around 0.6 whereas the minimum reached by the Q2D is near 0.5. Nonetheless, if the averaged efficiency is compared, as in Figure 2.18, the discrepancies between models diminish. Indeed, the expanded uncertainty of this measurement (with a coverage factor of 3 and in red in Figure 2.18) is approximately in the order of magnitude of the difference presented between the different models. Thus, with the available data and results, it is not possible to affirm that the quasi-2D model better captures the instantaneous efficiency and it would be risky to obtain any trend or conclusion. For future experimental campaigns, it will be crucial to obtain the instantaneous efficiency and determine the difference between models in terms of instantaneous efficiency prediction. Finally, it is worth noticing that the difference between the maximum and minimum instantaneous efficiency pulse is somewhere around 5-10 % depending on the case, which is high enough to affect the global performance of the engine and gives a good indication of the relevance of being able to predict the pulsating

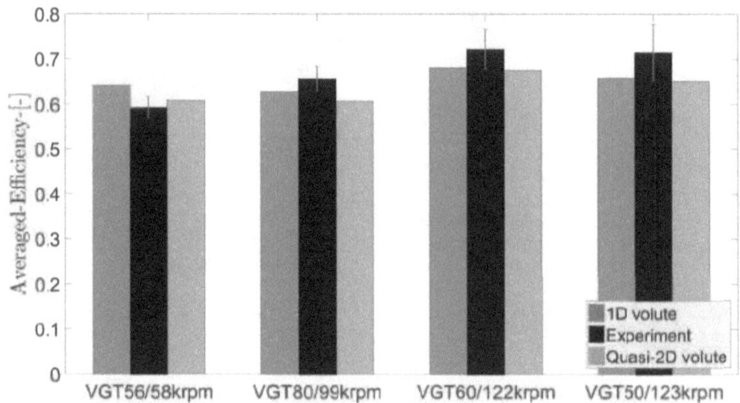

Figure 2.18: Turbine pulse-average efficiency.

flow conditions.

2.6 Conclusions

In this chapter, a method to measure the pressure of a radial turbine under continuous and pulsating flow conditions is presented. The experimental results obtained are used to fit and validate a quasi-two-dimensional model and to compare it with the classic one-dimensional model.

The pulsating flow conditions are generated by means of a rotating valve placed upstream of the turbine, designed in a way that would approximate the behaviour of that found in the exhaust manifold of a four cylinder, four strokes reciprocating engine. Furthermore, piezoelectric transducers are placed in the turbine inlet and outlet lines for beamforming purposes. That way, relevant experimental information about pulsating conditions in a wide range of turbocharged engine operating points has been obtained, in particular, incident and reflected waves for the turbine inlet and transmitted and second reflection waves for the turbine outlet can be analysed.

The quasi-two-dimensional model has proven to be an appropriate method to improve the previous one-dimensional duct approaches, especially at high frequencies. The classic volute model produces good predictions at up to 1000 Hz, while the new model improves these results and obtained better predictions at up to 2000 Hz. Contrary to what occurs in the fully one-dimensional model, in the quasi-two-dimensional volute different stator channels are exposed to different inlet condition. In this manner, the volute produces an averaging effect in the pressure pulse at high frequencies that reduces the amplitude of the signal and can't be reproduced using a one-dimensional duct. Higher frequencies remain out of the scope of this study, as they are affected by flow phenomena that is neither simulated nor modelled, so it could only be properly reproduced with further development of the model. This improvement in the accuracy of the results comes with a slight increment of the computational cost, in particular, the additional source term must be computed for each volute cell and around twice the number of volute cells are used so, as a result, for the same cell size, the quasi-2D approach doubles the time while computing the volute element.

It is worth mentioning that a better prediction of the conditions in the turbine line can be crucial for obtaining a more optimised design of the silencer and aftertreatment systems. This affects the pumping losses so the total fuel consumption and the pollutant emissions will be affected as well. Furthermore, a different design implies different back pressure and, in consequence, different turbocharger matching.

Nowadays, one-dimensional and zero-dimensional models of full engines are fast enough to be implemented in real-time simulations such as hardware-in-the-loop (HIL) with an acceptable but also improvable precision. Thus, in the following years, maintaining low computational costs in the turbocharger code will be crucial, specially considering that the development and the application of these systems are becoming more and more essential in the engine research field. Furthermore, the quasi-two-dimensional model should be also trivially adapted to twin-entry and waste-gate turbines, such as the ones used in petrol engines.

Chapter 2 References

[17]   CMT – Motores T'ermicos, Universitat Politècnica de València. *OpenWAM*. 2016. URL: http://www.openwam.org/ (cit. on pp. 7, 11, 22, 98, 133, 178).

[21]   E. Toro, M. Spruce, and W. Speares. "Restoration of the contact surface in the HLL-Riemann solver". English. *Shock Waves* 4(1) (1994), pp. 25–34. ISSN: 0938-1287. DOI: 10.1007/BF01414629 (cit. on pp. 8, 23).

[24]   H. Chen and D. Winterbone. "A method to predict performance of vaneless radial turbine under steady and unsteady flow conditions". In: *Turbocharging and Turbochargers*. Institution of Mechanical Engineers. 1990, pp. 13–22 (cit. on pp. 8, 20, 21, 97).

[25]   J. R. Serrano, F. J. Arnau, V. Dolz, A. Tiseira, and C. Cervelló. "A model of turbocharger radial turbines appropriate to be used in zero- and one-dimensional gas dynamics codes for internal combustion engines modelling". *Energy Conversion and Management* 49(12) (2008), pp. 3729 –3745. ISSN: 0196-8904. DOI: 10.1016/j.enconman.2008.06.031 (cit. on pp. 8, 30, 31, 96, 99, 100, 134).

[27] F. Payri, P. Olmeda, F. J. Arnau, A. Dombrovsky, and L. Smith. "External heat losses in small turbochargers: Model and experiments". *Energy* 71 (2014), pp. 534 –546. ISSN: 0360-5442. DOI: 10.1016/j.energy.2014.04.096 (cit. on pp. 8, 20, 27, 92, 131).

[28] J. R. Serrano, P. Olmeda, F. J. Arnau, A. Dombrovsky, and L. Smith. "Turbocharger heat transfer and mechanical losses influence in predicting engines performance by using one-dimensional simulation codes". *Energy* 86 (2015), pp. 204 –218. DOI: 10.1016/j.energy.2015.03.130 (cit. on pp. 8, 27, 92, 131).

[29] A. Gil, A. Tiseira, L. M. García-Cuevas, T. Rodríguez Usaquén, and G. Mijotte. "Fast three-dimensional heat transfer model for computing internal temperatures in the bearing housing of automotive turbochargers". *International Journal of Engine Research* (2018). DOI: 10.1177/1468087418804949 (cit. on pp. 8, 27, 92, 131).

[30] J. R. Serrano, F. J. Arnau, L. M. García-Cuevas, A. Dombrovsky, and H. Tartoussi. "Development and validation of a radial turbine efficiency and mass flow model at design and off-design conditions". *Energy Conversion and Management* 128 (2016), pp. 281 –293. ISSN: 0196-8904. DOI: 10.1016/j.enconman.2016.09.032 (cit. on pp. 8, 26–29, 96, 98, 131, 134, 135, 139, 157, 178).

[32] J. R. Serrano, P. Olmeda, A. Tiseira, L. M. García-Cuevas, and A. Lefebvre. "Theoretical and experimental study of mechanical losses in automotive turbochargers". *Energy* 55(0) (2013), pp. 888 –898. ISSN: 0360-5442. DOI: 10.1016/j.energy.2013.04.042 (cit. on pp. 8, 21, 92, 131).

[35] A. W. Costall, R. M. McDavid, R. F. Martínez-Botas, and N. C. Baines. "Pulse performance modelling of a twin-entry turbocharger turbine under full unequal admission". In: *Proceedings of ASME Turbo Expo 2009*. 2009. ASME, 2009. DOI: 10.1115/1.4000566 (cit. on pp. 8, 20, 130).

[39] M. S. Chiong, S. Rajoo, A. Romagnoli, A. W. Costall, and R. F. Martinez-Botas. "Assessment of Partial-Admission Characteristics in Twin-Entry Turbine Pulse Performance Modelling". In: *Volume 2C: Turbomachinery*. 2015, V02CT42A022. ISBN: 978-0-7918-5665-9. DOI: 10.1115/GT2015-42687 (cit. on pp. 8, 24, 92).

[60] U. Kesgin. "Effect of turbocharging system on the performance of a natural gas engine". *Energy Conversion and Management* 46(1) (2005), pp. 11 –32. ISSN: 0196-8904. DOI: 10.1016/j.enconman.2004.02.006 (cit. on p. 20).

[61] H. Tang, A. Pennycott, S. Akehurst, and C. J. Brace. "A review of the application of variable geometry turbines to the downsized gasoline engine". *International Journal of Engine Research* 16(6) (2015), pp. 810–825. DOI: 10.1177/1468087414552289 (cit. on p. 20).

[62] A. Pesiridis. "The application of active control for turbocharger turbines". *International Journal of Engine Research* 13(4) (2012), pp. 385–398. DOI: 10.1177/1468087411435205 (cit. on p. 20).

[63] A. Romagnoli and R. Martinez-Botas. "Performance prediction of a nozzled and nozzleless mixed-flow turbine in steady conditions". *International Journal of Mechanical Sciences* 53(8) (2011), pp. 557 –574. ISSN: 0020-7403. DOI: 10.1016/j.ijmecsci.2011.05.003 (cit. on p. 20).

[64] F. Payri, J. R. Serrano, P. Fajardo, M. A. Reyes-Belmonte, and R. Gozalbo-Belles. "A physically based methodology to extrapolate performance maps of radial turbines". *Energy Conversion and Management* 55(0) (2012), pp. 149 – 163. ISSN: 0196-8904. DOI: 10.1016/j.enconman.2011.11.003 (cit. on p. 20).

[65] V. De Bellis and S. Marelli. "One-dimensional simulations and experimental analysis of a wastegated turbine for automotive engines under unsteady flow conditions". *Proceedings of the Institution of Mechanical Engineers, Part D: Journal of Automobile Engineering* 229(13) (2015), pp. 1801–1816. DOI: 10.1177/0954407015571672 (cit. on p. 20).

[66] D. Bohn, N. Moritz, and M. Wolff. "Conjugate Flow and Heat Transfer Investigation of a Turbo Charger: Part II — Experimental Results". *ASME Conference Proceedings* 2003(3686) (2003), pp. 723–729. DOI: 10.1115/GT2003-38449 (cit. on p. 20).

[67] J. R. Serrano, P. Olmeda, A. Páez, and F. Vidal. "An experimental procedure to determine heat transfer properties of turbochargers". *Measurement Science and Technology* 21(3) (2010), p. 035109. DOI: 10.1088/0957-0233/21/3/035109 (cit. on pp. 20, 34, 95).

[68] J. Serrano, P. Olmeda, F. Arnau, and A. Dombrovsky. "General Procedure for the Determination of Heat Transfer Properties in Small Automotive Turbochargers". *SAE International Journal of Engines* 8(1) (2014), pp. 2014-01-2857. ISSN: 1946-3944. DOI: 10.4271/2014-01-2857 (cit. on p. 20).

[69] J. R. Serrano, P. Olmeda, F. J. Arnau, and V. Samala. "A holistic methodology to correct heat transfer and bearing friction losses from hot turbocharger maps in order to obtain adiabatic efficiency of the turbomachinery". *International Journal of Engine Research* (2019), p. 146808741983419. ISSN: 1468-0874. DOI: 10.1177/1468087419834194 (cit. on p. 20).

[70] R. Burke, C. Copeland, T. Duda, and M. Reyes-Belmote. "Lumped capacitance and three-dimensional computational fluid dynamics conjugate heat transfer modeling of an automotive turbocharger". *Journal of Engineering for Gas Turbines and Power* 138(9) (2016). DOI: 10.1115/1.4032663 (cit. on p. 20).

[71] X. Gao, B. Savic, and R. Baar. "A numerical procedure to model heat transfer in radial turbines for automotive engines". *Applied Thermal Engineering* (2019). ISSN: 13594311. DOI: 10.1016/j.applthermaleng.2019.03.014 (cit. on p. 20).

[72] P. Olmeda, V. Dolz, F. J. Arnau, and M. A. Reyes-Belmonte. "Determination of heat flows inside turbochargers by means of a one dimensional lumped model". *Mathematical and Computer Modelling* 57(7-8) (2013), pp. 1847 –1852. ISSN: 0895-7177. DOI: 10.1016/j.mcm.2011.11.078 (cit. on p. 21).

[73] J. R. Serrano, P. Olmeda, F. J. Arnau, M. A. Reyes-Belmonte, and A. Lefebvre. "Importance of Heat Transfer Phenomena in Small Turbochargers for Passenger Car Applications". *SAE Int. J. Engines* 6(2) (2 2013), pp. 716 –728. DOI: 10. 4271/2013-01-0576 (cit. on p. 21).

[74] S. Marelli, G. Marmorato, and M. Capobianco. "Evaluation of heat transfer effects in small turbochargers by theoretical model and its experimental validation". *Energy* 112 (2016). ISSN: 03605442. DOI: 10.1016/j.energy.2016.06.067 (cit. on p. 21).

[75] A. Romagnoli, A. Manivannan, S. Rajoo, M. S. Chiong, A. Feneley, A. Pesiridis, and R. F. Martinez-Botas. *A review of heat transfer in turbochargers*. 2017. DOI: 10.1016/j.rser.2017.04.119 (cit. on p. 21).

[76] H. Aghaali, H.-E. Ångström, and J. R. Serrano. "Evaluation of different heat transfer conditions on an automotive turbocharger". *International Journal of Engine Research* 16(2) (2015), pp. 137–151. DOI: 10.1177/1468087414524755 (cit. on p. 21).

[77] S. Marelli, S. Gandolfi, and M. Capobianco. *Experimental and Numerical Analysis of Mechanical Friction Losses in Automotive Turbochargers*. SAE Technical Paper 2016-01-1026. SAE International, 2016. DOI: 10.4271/2016-01-1026 (cit. on p. 21).

[78] J. R. Serrano, P. Olmeda, A. Tiseira, L. M. García-Cuevas, and A. Lefebvre. "Importance of Mechanical Losses Modeling in the Performance Prediction of Radial Turbochargers under Pulsating Flow Conditions". *SAE Int. J. Engines* 6(2) (2 2013), pp. 729 –738. DOI: 10.4271/2013-01-0577 (cit. on p. 21).

[79] J. Galindo, P. Fajardo, R. Navarro, and L. M. García-Cuevas. "Characterization of a radial turbocharger turbine in pulsating flow by means of CFD and its application to engine modeling". *Applied Energy* 103(0) (2013), pp. 116 –127. ISSN: 0306-2619. DOI: 10.1016/j.apenergy.2012.09.013 (cit. on pp. 21, 56, 90, 130).

[80] J. Galindo, S. Hoyas, P. Fajardo, and R. Navarro. "Set-up analysis and optimiza-tion of CFD simulations for radial turbines". *Engineering Applications of Com-putational Fluid Mechanics* 7(4) (2013), pp. 441–460. DOI: 10.1080/19942060.2013.11015484 (cit. on p. 21).

[81] I Hakeem, C.-C. Su, A. Costall, and R. F. Martínez-Botas. "Effect of volute geometry on the steady and unsteady performance of mixed-flow turbines". In: *Proceedings of the Institution of Mechanical Engineers Part A-Journal of Power and Energy*. Vol. 221. 2007, pp. 535–550. DOI: 10.1243/09576509JPE314 (cit. on p. 21).

[82] X. Hu. "An advanced turbocharger model for the internal combustion engine". PhD book. Purdue University, 2000 (cit. on p. 21).

[83] A. King. "A turbocharger unsteady performance model for the GT-Power internal combustion engine simulation". PhD book. Purdue University, 2002 (cit. on p. 21).

[84] A. Feneley, A. Pesiridis, and H. Chen. "A one-dimensional gas dynamics code for turbocharger turbine pulsating flow performance modelling". In: *Proceedings of the ASME Turbo Expo*. Vol. 8. American Society of Mechanical Engineers (ASME), 2017. ISBN: 9780791850954. DOI: 10.1115/GT2017-64743 (cit. on p. 21).

[85] S. Rajoo and M.-B. Ricardo. "Variable Geometry Mixed Flow Turbine for Turbochargers: An Experimental Study". *International Journal of Fluid Machinery and Systems* 1(1) (Oct. 2008), pp. 155–168. DOI: 10.5293/IJFMS.2008.1.1.155 (cit. on pp. 21, 57).

[86] K. G. Hohenberg, P. J. Newton, R. F. Martinez-Botas, M. Halamek, K. Maeda, and J. Bouilly. "Development and Experimental Validation of a Low Order Turbine Model Under Highly Pulsating Flow". In: *Turbo Expo: Power for Land, Sea, and Air, Volume 2C: Turbomachinery*. ASME. June 2017, V02CT44A023. DOI: 10.1115/GT2017-63983 (cit. on pp. 21, 57, 90, 130).

[87] J. R. Serrano, A. Tiseira, L. M. García-Cuevas, L. Inhestern, and H. Tartoussi. "Radial turbine performance measurement under extreme off-design conditions". *Energy* 125 (2017), pp. 72–84. DOI: 10.1016/j.energy.2017.02.118 (cit. on pp. 21, 57, 90, 135).

[88] A. J. Torregrosa, A. Broatch, R. Navarro, and J. García-Tíscar. "Acoustic characterization of automotive turbocompressors". *International Journal of Engine Research* 16(1) (2015), pp. 31–37. DOI: 10.1177/1468087414562866 (cit. on pp. 21, 57, 90, 130).

[89] O. Leufvén and L. Eriksson. "Measurement, analysis and modeling of centrifugal compressor flow for low pressure ratios". *International Journal of Engine Research* 17(2) (2016), pp. 153–168. DOI: 10.1177/1468087414562456 (cit. on p. 21).

[90] J. Galindo, A. Tiseira, R. Navarro, D. Tarí, and C. Meano. "Effect of the inlet geometry on performance, surge margin and noise emission of an automotive turbocharger compressor". *Applied Thermal Engineering* 110 (2017), pp. 875–882. DOI: 10.1016/j.applthermaleng.2016.08.099 (cit. on pp. 21, 130).

[91] J. Galindo, A. Tiseira, P. Fajardo, and L. M. García-Cuevas. "Development and validation of a radial variable geometry turbine model for transient pulsating flow applications". *Energy Conversion and Management* 85 (2014), pp. 190–203. ISSN: 0196-8904. DOI: 10.1016/j.enconman.2014.05.072 (cit. on pp. 21, 25, 131).

[92] Z. Ding, W. Zhuge, Y. Zhang, H. Chen, and R. Martinez-Botas. "Investigation on pulsating flow effect of a turbocharger turbine". In: *American Society of Mechanical Engineers, Fluids Engineering Division (Publication) FEDSM*. Vol. 1A-2017. 2017. ISBN: 9780791858042. DOI: 10.1115/FEDSM2017-69186 (cit. on p. 21).

[93] Z. Ding, W. Zhuge, Y. Zhang, H. Chen, R. Martinez-Botas, and M. Yang. "A one-dimensional unsteady performance model for turbocharger turbines". *Energy* (2017). ISSN: 0360-5442. DOI: 10.1016/j.energy.2017.04.154 (cit. on pp. 21, 130, 131).

[94] J. Galindo, J. R. Serrano, F. J. Arnau, and P. Piqueras. "Description of a Semi-Independent Time Discretization Methodology for a One-Dimensional Gas Dynamics Model". *Journal of Engineering for Gas Turbines and Power* 131(3) (2009), p. 034504. ISSN: 07424795. DOI: 10.1115/1.2983015 (cit. on pp. 23, 133).

[95] S. K. Godunov. "A Difference Scheme for Numerical Solution of Discontinuous Solution of Hydrodynamic Equations". *Matematicheskii Sbornik* 47 (1959), pp. 271–306 (cit. on pp. 23, 178).

[96] B. van Leer. "Towards the Ultimate Conservation Difference Scheme. II. Monotonicity and Conservation Combined in a Second-Order Scheme". *Journal of Computational Physics* 14 (Mar. 1974), p. 361. DOI: 10.1016/0021-9991(74)90019-9 (cit. on p. 23).

[97] R. Courant, K. Friedrichs, and H. Lewy. "Über die partiellen Differenzengleichungen der mathematischen Physik". *Mathematische Annalen* 100(1) (Dec. 1928), pp. 32–74. ISSN: 0025-5831. DOI: 10.1007/bf01448839 (cit. on p. 24).

[98] J. R. Serrano, F. J. Arnau, L. M. García-Cuevas, and L. B. Inhestern. "An innovative losses model for efficiency map fitting of vaneless and variable vaned radial turbines extrapolating towards extreme off-design conditions". *Energy* 180 (2019), 626 –639. ISSN: 0360-5442. DOI: 10.1016/j.energy.2019.05.062 (cit. on pp. 26, 157).

[99] *Supercharger Testing Standard.* SAE J1723. Society of Automotive Engineers, 1995 (cit. on p. 33).

[100] *Turbocharger gas stand test code.* SAE J1826. Society of Automotive Engineers, 1995 (cit. on p. 33).

[101] G Pinero, L Vergara, J. M. Desantes, and A Broatch. "Estimation of velocity fluctuation in internal combustion engine exhaust systems through beamforming techniques". *Measurement Science and Technology* 11(11) (2000), p. 1585. DOI: 10.1088/0957-0233/11/11/307 (cit. on pp. 36, 39, 64, 133, 177).

[102] P. Welch. "The Use of Fast Fourier Transform for the Estimation of Power Spectra: A Method Based on Time Averaging Over Short, Modified Periodograms". *IEEE Transactions on Audio Electroacoustics* 15 (2 1967), pp. 70–73. ISSN: 0018-9278. DOI: 10.1109/TAU.1967.1161901 (cit. on pp. 39, 106, 136).

[103] F. Harris. "On the use of windows for harmonic analysis with the discrete Fourier transform". In: *Proceedings of the IEEE.* Vol. 66. IEEE, 1978, pp. 51–83. DOI: 10.1109/PROC.1978.10837 (cit. on pp. 39, 106).

Chapter 3

An experimental method to test two-scroll turbines under non-linear pulse conditions

Contents

3.1 Introduction

For the past few years, the automotive industry has been living a deep and accelerated evolution. Gasoline engines are gradually gaining importance and market share against diesel engines, even in Europe, where the industry is facing the constant challenge of improving the engine efficiency and decreasing the pollutants emissions, following the path drawn by the EURO 6 and the future EURO 7 regulations [104]. Furthermore, electrical engines are increasing its market share but, however, as it is exposed in [6], internal combustion engines have still a long way to go. In this context, the widespread use of the turbocharging technology allows more efficient and downsized engines with considerable higher specific power. Thus, reducing engine weight and mechanical losses, which leads to a drastic reduction of CO_2 among other pollutants [105].

Vehicle emissions and fuel consumption reduction are not, however, the only concern of automotive manufacturers. The customer's requirements concerning the driving experience is forcing the industry to develop more and more evolved techniques in the field of turbocharging and, hence, increasing the complexity of the system. As an example, dual-turbocharger technology is nowadays considerably present in some applications [106], while twin-entry and double-entry turbines are being selected more frequently in the last high efficiency engines.

Twin and double entry turbines are characterised by the particular design of their housing, in which the flow is divided into two separated flow channels fed by separate exhaust banks, decoupling the energy pulses of sequentially igniting cylinders. Due to the avoidance of backflows generated by interferences of exhaust process between consecutive firing order cylinders, a better engine volumetric efficiency is obtained. Furthermore, a more effective use of the dynamic pressure of the gas pulses improves significantly the transient response compared to a single entry turbine, particularly evident when comparing via engine Worldwide harmonized Light vehicles Test Cycle (WLTC) [41].

Two-scroll turbines can be classified in two main groups, the twin-scroll turbines and the double-scroll turbines, depending on how the volute is designed; twin-scroll if the housing of the volute is meridionally divided, double-scroll if the division of the volute housing is circumferential. Many studies have been carried out with the intention of analysing the differences in performance of twin- and double-volute designs, especially in terms of efficiency, like the investigation developed by Pischinger and Wunsche in [107] or by Romagnoli et al. in [37]. However, the objective of this chapter is to perform a comparison or analysis of a particular two-scroll turbine design, since the experimental method that is explained is flexible enough to be used for any turbine design.

The experimental data provided by the turbocharger manufacturers is usually obtained in normal-steady conditions. Nevertheless, during the actual driving conditions, the turbine is continually hit by highly pulsating flow due to the opening and closing of the cylinder valves, as explained in [79]. Thus, a good comprehension of the behaviour of the turbine under pulsating conditions has become very useful for the automotive manufacturers during the design stage. To satisfy this demand, several experiments have been developed for single-entry turbines, as the works carried out by Hohenberg et al. [86] for pulsating flow in turbines or Serrano et al. [87] and Torregrosa et al. [88] for the performance and off-design conditions among many other studies. Especially focused on the turbocompressor performance under pulsating conditions for diesel engines is the experimental investigation carried out by Torregrosa et al [108]. A gas stand with a modular design philosophy is exposed in [109] that provides a lot of flexibility when the aim is to analyse the response of very different pulses. Finally, a similar methodology as the one described here can be consulted in [1], being, in this case, applicable only for single-entry turbine configurations.

Regarding the unsteady behaviour in the twin-entry and double-entry turbines, more recent investigations have been carried out. First, Arcoumanis et al. [110] and later Rajoo et al. [85] investigated the influence of the pulsation on a mixed-flow turbine and the response of different configurations. On the other hand, a comparison of the performance of a double-entry turbine under pulsating and steady conditions can be found in [111]. More recent is the study carried out by Mingyang et al. [112], where, apart from an analytical model of the unsteadiness of the mixed-flow turbine, a 1-D unsteady model is validated against experimental results. Due to the challenge that obtaining reliable pulsating experimental measurements implies, many studies have been developed using CFD simulations, as the works in [42, 113, 114] and more recently in [115]. Finally, in [116] a deep CFD numerical analysis for twin turbine is presented along with an experimental validation with a wide range of operating conditions.

Particularly interesting is the work developed by Kabral et al in [117], where a first acoustic characterisation is performed by means of pressure decomposition, obtaining the small amplitude (linear) dynamic response of the twin-scroll turbine. The present investigation adds, however, the capability to test any two-scroll turbine under higher amplitudes with the aim of obtaining a more realistic acoustic characterization of the turbine. This provides enough flexibility to check the dependency of the response to different excitations in hot and cold conditions, different amplitudes and different engine points of operation. It is also possible to separate the effects of each branch independently. This experimental data provide very valuable information of the nature of the pulse reflected to the exhaust manifold and transmitted to the exhaust line, allowing more accurate decisions during the design and matching stages and optimising the scavenging effects and the high efficiency turbine range. High quality experimental data is also essential for developing and validating reliable 1D models, in fact, by means of the procedure described in this chapter, it has been possible to obtain the experimental data used to perform the model validation presented in the next chapter.

3.2   Experimental method description

The experimental work that will be here detailed has been carried out in the CMT-Motores Térmicos gas stand laboratory for turbocharging research, an especially built facility for testing any type of turbocharger in any condition. In this installation the turbocharger is isolated from the IC engine and measured under controlled conditions. In this way, the gas stand configuration provides the greater range of possibilities, including steady and pulsating flow conditions, hot and cold flow conditions, as well as amplitude and frequency control and heat fluxes characterisation, all with complete flexibility and minimum changes in the installation.

Firstly, the test bench setup is described in detail and sketched, then, the turbine characteristics are commented. Finally, the procedure for performing the pulsating flow measurements is described along with the necessary instrumentation and the test plan.

3.2.1   Test bench setup

Figure 3.1 represents the basis layout of the experimental gas stand, which can be divided into 4 main systems; the compressor line, the turbine line, the coolant circuit and the lubrication circuit.

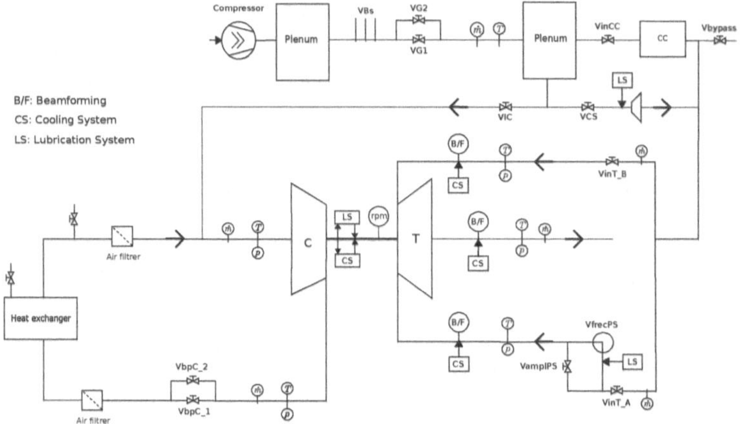

Figure 3.1: Testbench layout.

In the next table, a summary of the main sensors and the different elements of measure are presented.

Variable	Sensor type	Range	Symbol in the sketch
Gas pressure (mean)	Piezoresistive	0 to 5 bar	p
Gas pressure (instantaneous)	Piezoelectric	0 to 250 bar	B/F
Gas/metal temperature	K-type thermocouple	273 K to 1500 K	T
Gas mass flow	V-cone, thermal, vortex	45 to 1230 kg/h	$\dot{m}$
Oil pressure	Piezoresistive	0 to 5 bar	LS
Oil temperature	RTD	173 K to 723 K	LS
Oil mass flow	Coriolis	Few tens gr/s	LS
Turbocharger speed	Inductive sensor	< 300 krpm	rpm

Table 3.1: List of measurement equipment with corresponding symbols in the layout sketch.

3.2.1.1 Compressor line

In open loop, the compressor sucks air directly from the atmosphere of the test-bench and passes it first through a filter. Then, the mass flow rate is measured using a thermal mass flow meter. Downstream of the compressor, piezoresistive sensors and thermocouples are used for the measurement of the pressure and temperature respectively, along with a vortex-type mass flow meter for the compressor outlet mass flow. After taking the measurements, there is an electronically driven back pressure valve allowing to control the full working range for the compressor, from surge to choke. Finally, another filter is used to clean the air from oil before discharging it to the atmosphere.

The compressor can also be tested in a closed-loop configuration, where the outlet of the loop can be attached to its inlet. Thus, the compressor can be tested with low-pressure or high pressure at its inlet, lowering or raising the power drawn for a given compression ratio and a given corrected speed.

3.2.1.2 Turbine line

The air used to feed the turbine proceeds from a two-stage radial compressor located in an adjoining room. This compressor has a maximum pressure handling capacity of

5 bar, a maximum power of 500 kW, an intermediate cooling system, and an operative point control system. When the required pressure or mass flow rate is lower than the minimum supplied by the compressor, electronic discharge valves are used. These valves are placed upstream and downstream of two high-volume plenums. The discarded flow is directly discharged to the atmosphere. The two-stage radial compressor has its own filters, so the air that enters the turbine is oil and dust-free. The high-volume plenums are used to stabilise the flow.

After the stabilisation plenums, a combustion chamber heats the air to the desired turbine inlet temperature, with a maximum reachable temperature of 1200 K for the maximum flow rate. This element is not used for all the tests, as some of the operating points are obtained in cold conditions.

Downstream of the combustion chamber, the flow is divided into the two turbine branches, where new measurements of mass flow are obtained by means of V-cone-type mass flow meters. Due to their reliability and accuracy even in dirt or pulsating conditions, these elements are particularly interesting for this application.

For generating the pulses, a rotating valve activated by an electric engine is situated in one of the inlet branches. This rotating valve was designed to withstand high pressures and temperatures, so it can be used to generate engine-like pulses. During the tests, measurements had to be obtained pulsating in both inlet branches, so it was necessary to find an easily installable solution compatible with the rest of the elements of the turbine line. With this purpose, two flexible pipes were installed after the control pipes, allowing a fast change of the pulsating branch by simply switching their connections, as shown in Figure 3.2.

Finally, a beamforming array of three piezoelectric sensors for the pressure decomposition is placed before the inlet maintaining a specific distance among each other, as it will be detailed in the pulsating measurements section. The average flow pressure and temperature is also measured, using piezoresistive sensors and K-type thermocouples.

In the outlet of the turbine, another beamforming array is used for the pressure decomposition, along with new measurements of mean pressure, temperature and mass flow, all following the same methodology as in the turbine inlets.

3.2.1.3   Coolant circuit

The piezoelectric sensors used for the measurements of the instantaneous pressure are especially sensitive to temperature changes. Thus, the pressure sensors require a cooling system that maintains the temperature constant so that the pressure measurements are not affected. For this purpose, the general cooling system of the building is used, maintaining a coolant bath at a constant temperature of 20 °C.

Figure 3.2: Flexible pipes for switching the pulsating branch.

3.2.1.4   Oil circuit

A controlled lubrication system is used for both the turbocharger and the rotating valve. A Coriolis-type flow meter is used to measure the oil mass flow. The oil pressure and temperature at both the inlet and outlet of the turbine, as well as the pressure at the inlet of the rotating valve, are measured. The oil flow is kept controlled by means of a recirculation circuit and a piloted valve. To control the oil temperature, the system uses an electrical heater.

Lubrication inlet and outlet temperatures are measured by means of low uncertainty platinum resistance temperature detectors. Periodic samples of oil can be taken from the independent lubrication system in order to characterise its properties (viscosity, density and specific heat capacity variations with temperature). This way, any change in oil properties due to oxidation can be monitored.

3.2.2   Turbine characteristics

The turbocharger used for the tests presented consists of a radial compressor and a twin entry turbine, with asymmetrical entry and wastegate, for applications in an automotive commercial gasoline engine. Regarding the nomenclature, the two different entries of the turbine will be denominated as *shroud* and *hub* due to its location inside the housing.

3.2.3   Pulsating flow measurements

It is quite challenging to obtain reliable experimental data when dealing with realistic pulsating conditions. The instantaneous pressure sensors and their corresponding amplifier are very sensitive to the electrical noise present in the test-bench, and the output signal generated should be treated and filtered properly. The know-how and experience obtained in [1] has been determinant for the quality of the results. The addition of a second turbine entry requires a fine operation of the control systems and valves. In the following sections, the necessary control systems and valves for performing the pulsating flow measurements are described, as well as the data acquisition equipment and the test plan followed.

3.2.3.1   Pulse generation and frequency control

For generating a highly pulsating flow, an especially designed rotating valve is set up at one of the inlet branches of the turbine, using the flexible pipe (see Figure 3.2) for switching the pulsating branch. The valve was built from a truck turbocompressor, and it is powered by an asynchronous motor. The frequency of the electrical motor can be controlled from 0 to 50 Hz, in order to reach the desired engine equivalent speed (Figure 3.3).

Figure 3.3: Rotating valve and synchronous motor.

In the turbine side, a plate with three equispaced orifices is placed in order to allow or limit the passing flow and impart the pulses into one of the inlet branches. In each complete rotation of the motor shaft (360°) three pulses are transmitted, producing a maximum frequency of 150 Hz. The different shapes of the plate holes can imitate the opening and closing duration of the actual cylinder valves (Figure 3.4).

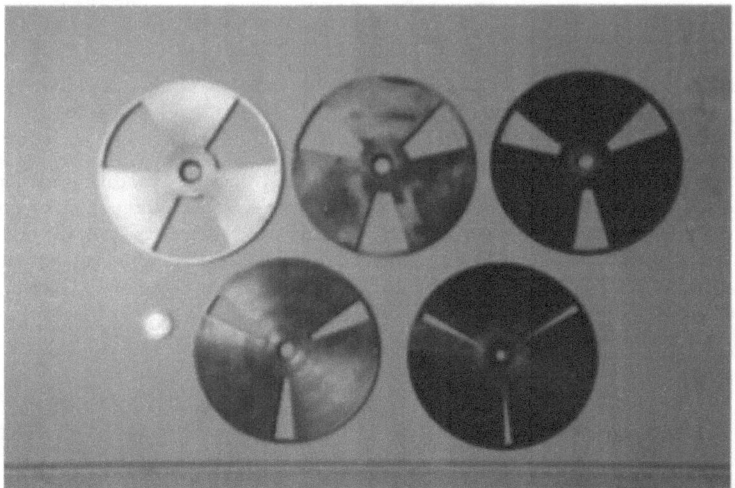

Figure 3.4: Rotating plates for pulse generation.

The frequency of the rotating valve can be easily adjusted by following a simple relationship between the electric motor frequency and the equivalent engine speed. For example, for performing tests for an equivalent engine speed of 2000 rpm:

$$2000 \cdot \frac{\text{rev}}{\text{min}} \cdot \frac{1\text{min}}{60\text{s}} \cdot \frac{4\text{cylinders}}{2\text{branches}} \cdot \frac{1}{3\text{holes}} = 11\,\text{Hz} \tag{3.1}$$

In this example, the frequency of the imposed pulse is 33 Hz, which requires a setpoint frequency for the electrical motor of 11 Hz, taking into account that each plate has three holes. The maximum equivalent engine speed available is, consequently, 9000 Hz.

3.2.3.2 Amplitude control

The amplitude of the generated pulse of each operating point is adjusted using the amplitude control valve operated by a PID controller. The valve operates by controlling the amount of mixing of the steady and pulsating flow, so that if the amount of steady mass flow is less in the mixing, the pulsating amplitude is higher and vice versa. This methodology allows to recreate the higher engine load operating points by simply increasing the pulsating amplitude. With this configuration, the rotating valve is fixed in one inlet, but, as it was mentioned before, the pulsating branch can be easily changed by switching the flexible pipe (Figure 3.2).

3.2.3.3   Admission control

The admission conditions were controlled by controlling the mass flow rate at each inlet. For setting the desired mass flow rate, the valve position was fixed before the experiment. Two different positions were used, half and full, in a strategy where the pulsating branch was always kept at full admission and the other branch at 50%. By changing the positions of the flexible pipe, partial admission conditions were either imposed at the shroud side entry or hub side entry of the turbine.

3.2.3.4   Pressure decomposition

For the instantaneous pressure, three pressure transducers were fixed in each inlet branch as well as in the outlet branch (Figure 3.5). For performing the pressure decomposition in each of the inlet branches and the outlet branch, the procedure followed [101] is analogous as the one already detailed in subsection 2.4.2. The only difference with the pressure decomposition description from chapter 2 is that there is an additional turbine inlet branch where a third piezoelectric sensor matrix should be placed.

To process the signal generated by the sensors, charge amplifiers were used. Every sensor was previously calibrated for adjusting the sensitivity. It is worth mentioning that piezoelectric sensors are highly sensitive to the electrical noise generated by all the equipment present in the test-bench. With the objective of lowering this noise as much as possible, all the sensors were connected to an equipotential earth by means of a low impedance metal mesh. On the other hand, piezoresistive sensors have been used for measuring the static mean pressure in the branches.

In addition, averaged temperature measures were also taken in the turbine inlet and outlet pipes using thermocouples. To compensate the natural uncertainty of this instruments as much as possible, four measures were taken in each section using four equispaced radial positions. The thermocouples are placed at different depths to get a more representative section average, as seen in Figure 3.6.

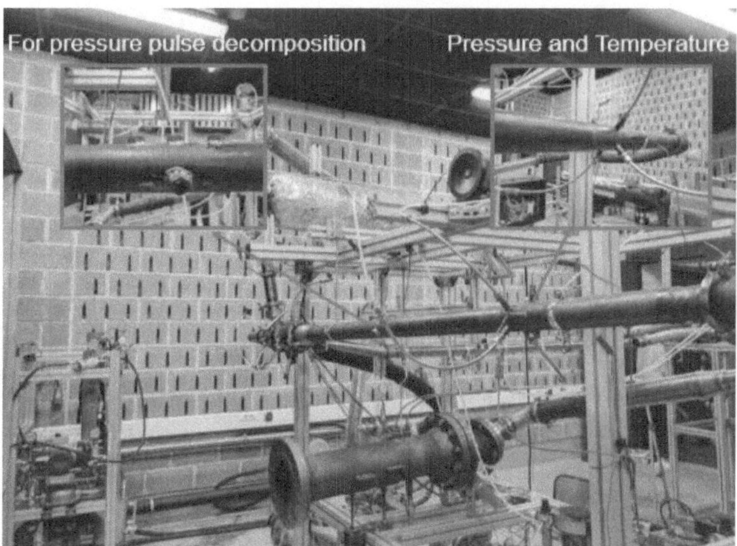

Figure 3.5: Instantaneous pressure sensors setup.

3.2.3.5 Control system and data acquisition

In this facility, the PID regulators are aimed to control the main regulation systems of the installation such as the two-stage radial compressor or the multiple valves. The digital PID that controls the compressor fixes its outlet pressure by modifying the rotational speed and the position of the variable geometry blades. All the valves in the installation are controlled by digital PIDs.

In Table 3.2, the valves that have been used in this experiment are enumerated. The last column contains the correspondent symbol of each valve correspondent with the sketch of Figure 3.1.

In an experimental campaign of this characteristics, not only a safe and reliable control is necessary, but a correct data acquisition system is also vital. This is particularly important if the purpose lies in obtaining high resolution unsteady flow measurements.

Therefore, two different compatible mechanisms were used, namely, the power regulator of each main element, and the specifically designed automatic system *full_rio*, which is an in-house manufactured equipment. This system is based on PXI and RIO modules of National Instruments, and it is operated by a software designed in LabVIEW environment.

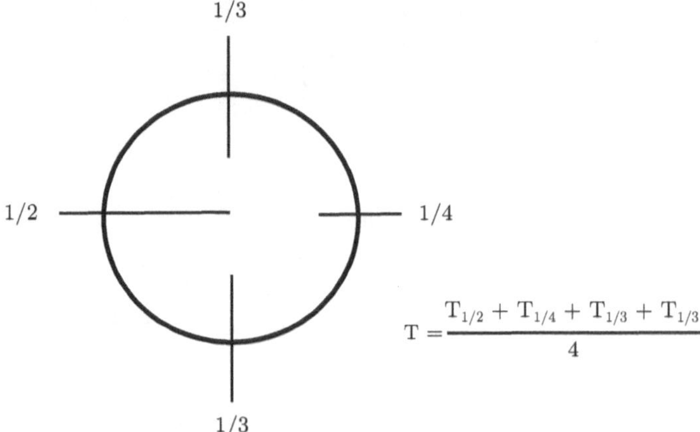

$$T = \frac{T_{1/2} + T_{1/4} + T_{1/3} + T_{1/3}}{4}$$

Figure 3.6: Thermocouples setup.

3.2.3.6   Test plan

Table 3.3 presents a summary of the working points selected for the experiments. The goal was to cover a full range of operative conditions, including different equivalent engine speed, flow conditions and compressor map operative range. The mass flow ratio (MFR) establishes a relation between the different mass flow quantities through each of the inlets, shroud and hub, and it is defined as in Equation 3.2:

$$MFR = \frac{\dot{m}_{\text{shroud}}}{\dot{m}_{\text{shroud}} + \dot{m}_{\text{hub}}} \tag{3.2}$$

3.3   Results and discussion

In the following section, first the results of the pressure decompositions measurements are exposed. Later, a complete analysis of the turbine behaviour under pulsating conditions is presented to provide a sample of the potential of these experiments when the goal is to study the turbine non-linear response.

Actuator	Action	Range	Symbol in the sketch
Vent valve	Open valve	0-100%	VBs
General valve (big)	Control experiment pressure and flow	0-100%	VG1
General valve (small)	Control experiment pressure and flow	0-100%	VG2
Combustion chamber valve	Control cold air	0-100%	VinCC
Cold air valve	Control cold air	0-100%	VCS
Compressor entry valve	Control compressor inlet mass flow	0-100%	VIC
Bypass valve	Control outlet combustion chamber	0-100%	Vbypass
Turbine entry A valve	Control valve through turbine inlet A	0-100%	VinT_A
Turbine entry B valve	Control valve through turbine inlet B	0-100%	VinT_B
Pulse amplitude valve	Control amplitude pulse	0-100%	VamplPS
Backpressure valve (big)	Control compressor backpressure	0-100%	VbpC_1
Backpressure valve (small)	Control compressor backpressure	0-100%	VbpC_2
Pulse Frequency valve	Control rotating valve frequency	0-100%	VfrecPS

Table 3.2: List of the control valves with corresponding symbols in the layout sketch.

3.3.1 Pressure decomposition measurements

The pressure decomposition results of the two operating points #4 and #46 (Table 3.3) in both time and frequency domain are presented from Figure 3.7 to Figure 3.9 and from Figure 3.10 to Figure 3.12 for both cases respectively. For the nomenclature, incident and reflected has been used in the pulsating branch for the forward and backward travelling waves, i.e., towards turbine and away from turbine respectively. Similar criterion has been used for the not pulsating inlet branch, where indirect reflected and indirect transmitted have been used for the waves travelling towards the turbine and away from the turbine respectively. On the other hand, the transmitted wave travels downstream

Case label	Pulsating branch	Mass flow ratio approx.	Flow conditions
#1 - #4	Shroud	0.5	Cold
#5 - #12	Shroud	0.7	Cold
#13 - #18	Hub	0.5	Cold
#19 - #26	Hub	0.3	Cold
#27 - #32	Hub	0.5	Cold
#33 - #40	Shroud	0.7	Hot
#41 - #48	Hub	0.3	Hot

Table 3.3: Summary of the pulsating flow test matrix for two-scroll turbine experiments.

of the turbine in the direction of the flow and the 2nd reflection wave travels from the outlet of the turbine in the opposite direction of the flow.

In the bottom graph of each plot the pressure decomposition results in the frequency domain are presented. The time domain original pressure measured is treated with a low-pass, linear phase filter and a Fourier transformation using Welch's periodogram method with a Hamming window. In this way, it is possible to directly observe the amplitude in dB of each harmonic of the equivalent engine. As it is expected, the amplitude decreases when the frequency is higher, going from 120-140 dB of the first harmonics to less than 100 dB for a frequency of 1000 Hz. The amplitude of the pulse in the high spectra is however, significant enough to not be neglected.

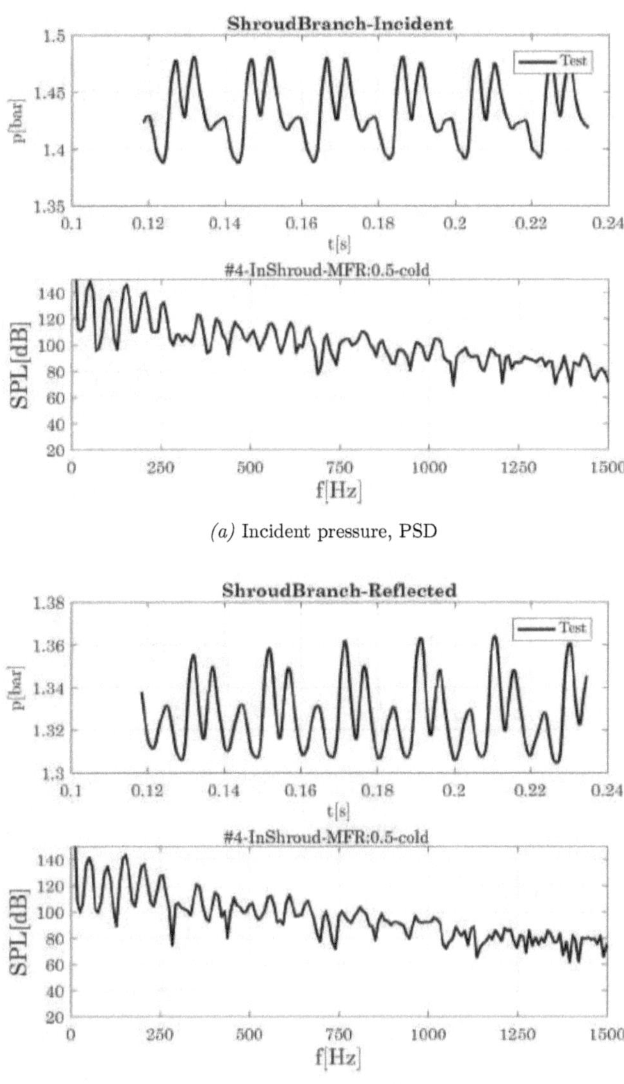

(a) Incident pressure, PSD

(b) Reflected pressure, PSD

Figure 3.7: (a), (b), Incident and reflected waves for case #4.

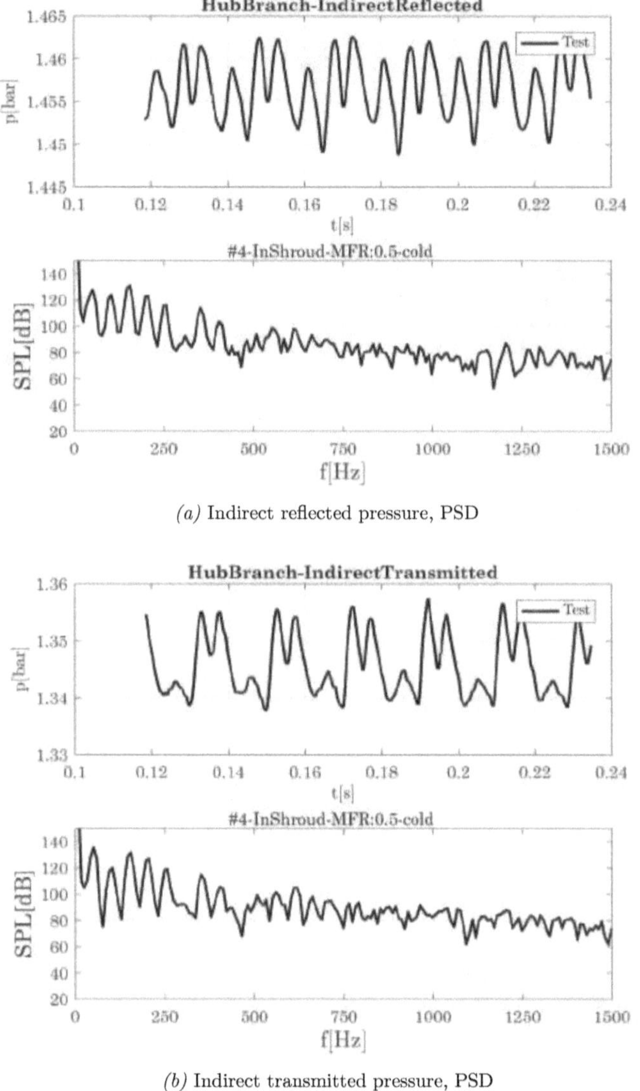

(a) Indirect reflected pressure, PSD

(b) Indirect transmitted pressure, PSD

Figure 3.8: (a), (b), Indirect reflected and indirect transmitted waves for case #4.

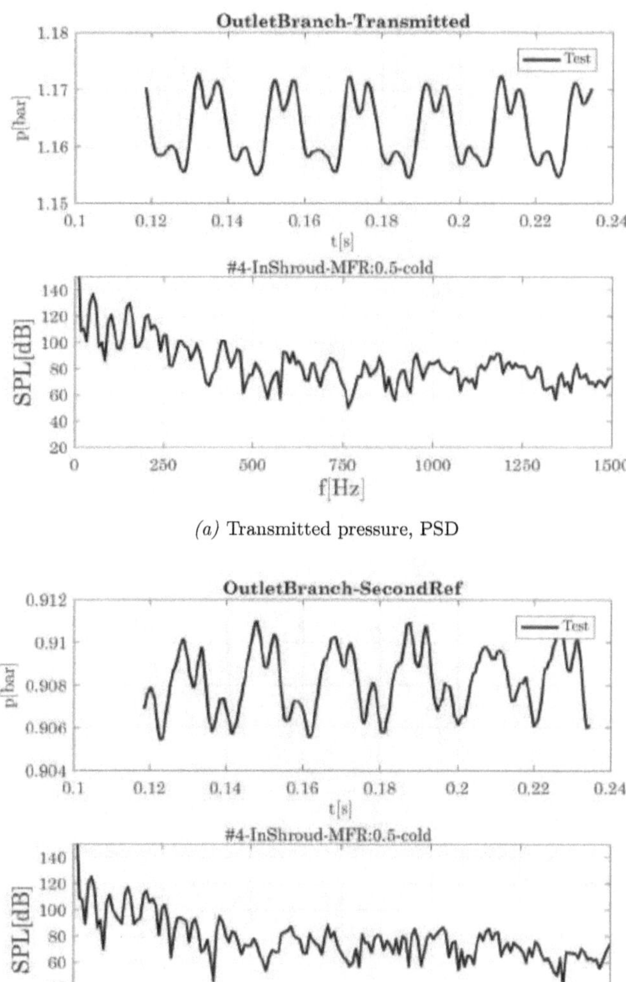

(a) Transmitted pressure, PSD

(b) 2nd reflection pressure, PSD

Figure 3.9: (a), (b), Transmitted and 2nd reflection for case #4.

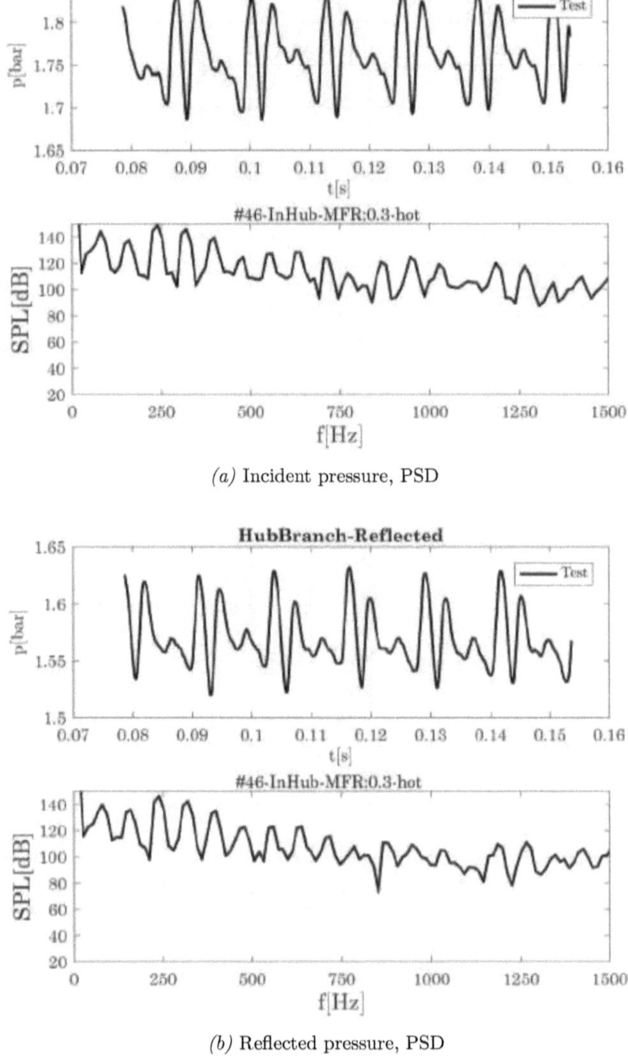

(a) Incident pressure, PSD

(b) Reflected pressure, PSD

Figure 3.10: (a), (b), Incident and reflected waves for case #46.

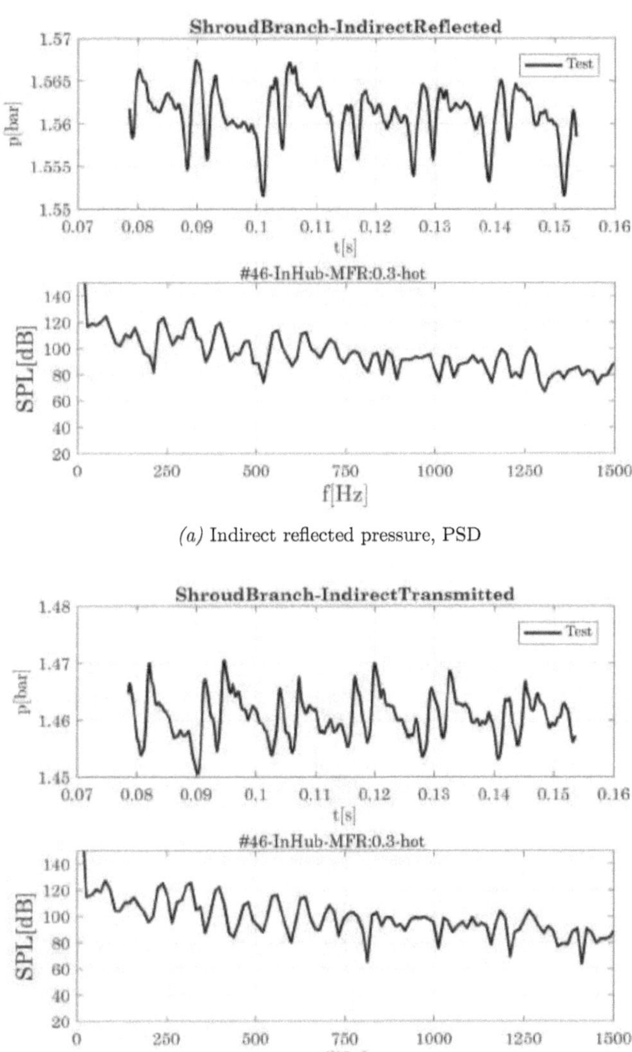

(a) Indirect reflected pressure, PSD

(b) Indirect transmitted pressure, PSD

Figure 3.11: (a), (b), Indirect reflected and indirect transmitted waves for case #46.

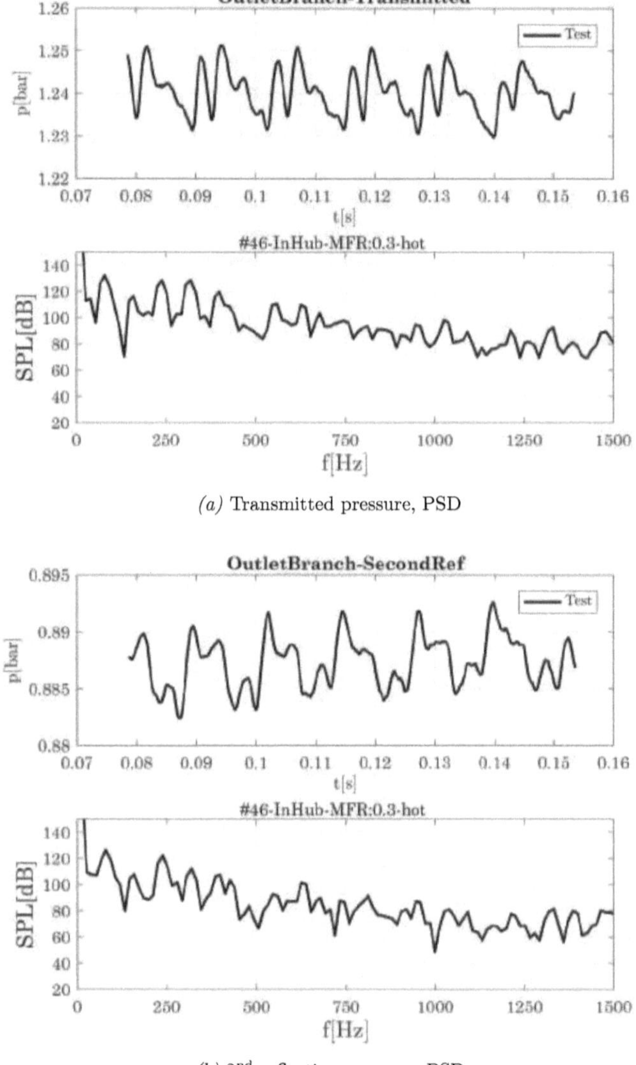

(a) Transmitted pressure, PSD

(b) 2nd reflection pressure, PSD

Figure 3.12: (a), (b), Transmitted and 2nd reflection for case #46.

The time domain graph provides relevant information of the shape and amplitude of the pulse and how it evolves when interacting with the turbine rotor. The reflected component (Figures 3.7(b) and 3.10(b)) is roughly conserving the shape of the incident, while its amplitude and mean level are clearly decreasing, a good indication that results are coherent.

Focusing in the transmitted wave of the hot flow operating point (Figure 3.12(a)), it can be observed how the mean level of the pressure pulse decreases from approximately 1.75 bar in the incident to 1.24 bar (transmitted) while the amplitude goes from more than 100 mbar to 20 mbar. In the #46 case, the pulses are performed in the hub branch, so in the indirect transmitted result (Figure 3.11(b)) what it can be observed is the effect in the shroud branch of the pulse that is produced in the hub. In this sense, it is interesting to analyse how the pulse evolves when it hits the rotor and travels backwards through the other inlet branch. It is important to mention that, although the amplitude of the indirect transmitted wave is considerably lower than the incident, it will have an impact in the performance and therefore, it should not be neglected.

3.3.2 Non-linearity assessment

In the linear-acoustics limit, with the turbine behaving as a linear passive element, Equation 3.3 should apply:

$$
\begin{pmatrix} R_1 & T_{2,1} & T_{3,1} \\ T_{1,2} & R_2 & T_{3,2} \\ T_{1,3} & T_{2,3} & R_3 \end{pmatrix} \cdot \begin{pmatrix} p^i_{1,+} & p^j_{1,+} & p^k_{1,+} \\ p^i_{2,+} & p^j_{2,+} & p^k_{2,+} \\ p^i_{3,-} & p^j_{3,-} & p^k_{3,-} \end{pmatrix} = \begin{pmatrix} p^i_{1,-} & p^j_{1,-} & p^k_{1,-} \\ p^i_{2,-} & p^j_{2,-} & p^k_{2,-} \\ p^i_{3,+} & p^j_{3,+} & p^k_{3,+} \end{pmatrix} \tag{3.3}
$$

where R_a is the reflection coefficient for branch a, $T_{a,b}$ is the transmission coefficient for a pulse going to the turbine from branch a and being transmitted to branch b, $p^i_{a,+}$ is the pressure wave travelling downstream in branch a and experiment number i and $p^i_{a,-}$ is the pressure wave travelling upstream in branch a and experiment number i. The system could be solved and the matrix of reflection and transmission coefficients obtained if a system of linearly-independent experiments were performed. Linearly-independent experiments can be performed as shown in [117], producing pulses with different excitations at both turbine inlets and the turbine outlet. When a single excitation source is used to generate active pulses in one of the branches, for example generating a pulse $p^i_{1,+}$, the values of $p^i_{2,+}$ and $p^i_{3,-}$ are also proportional to the amplitude of $p^i_{1,+}$ if the system is working in the linear acoustics limit. Thus, when operating the system at a constant average working point (i.e., constant average pressure ratio and reduced speed) and producing different pulses in the same turbine inlet, it is not possible to generate linearly-independent, linear acoustics experiments. Using data measured at a constant working point but different pulse shapes on the same inlet branch, the second matrix in Equation 3.3 was computed and its determinant was found to be different enough than 0 to indicate linear independence. Thus, the pulses were strong enough to induce non-linear acoustic behaviour in the turbine.

3.3.3 Turbine behaviour analysis

The goal of the following results is to exemplify firstly the potential of this kind of experimental study in terms of analysing the non-linear behaviour of the engine pulses in both turbine inlets and the turbine outlet. Furthermore, presenting a summary of all the operating points in the same graph allows to obtain different conclusions that can be very valuable.

In Figure 3.13, Figure 3.14 and Figure 3.15, the reflected, indirect transmitted and transmitted coefficients are presented for all the cases tested versus the mass flow ratio. These coefficients are simply obtained as a quotient between the maximum amplitude of the reflected, indirect transmitted or transmitted components and the maximum amplitude of the incident component.

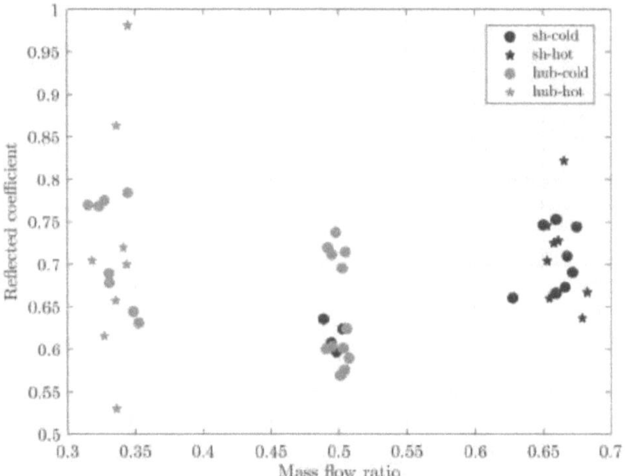

Figure 3.13: Reflected coefficient for all the operating points.

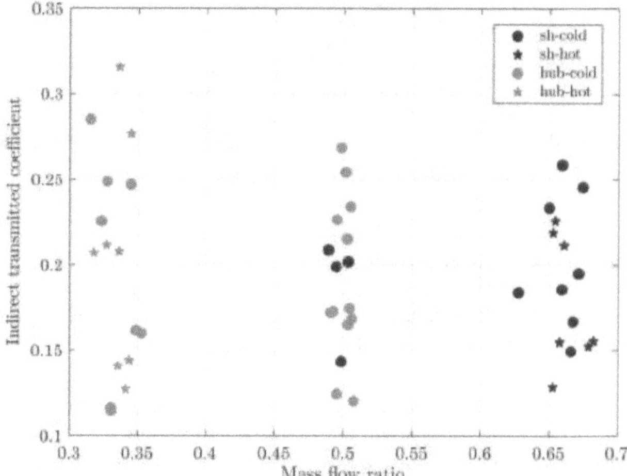

Figure 3.14: Indirect transmitted coefficient for all the operating points.

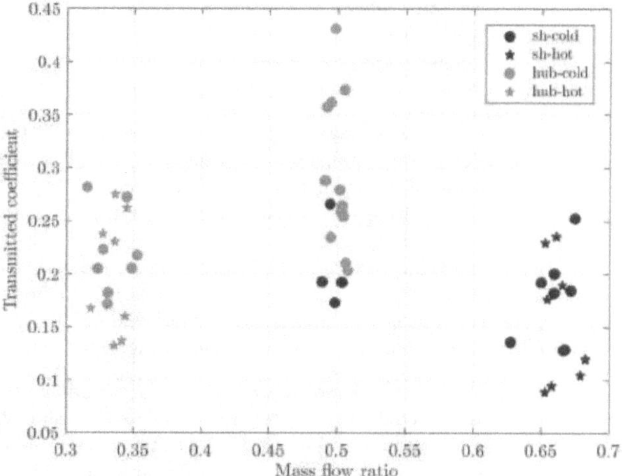

Figure 3.15: Transmitted coefficient for all the operating points.

In this study, three main conclusions can be inferred:

- The hot flow operating points are of course higher in amplitude than the cold cases. Nevertheless, when normalising with the incident component, it is clear that the coefficients of hot and cold points are moving in the same range, and, thus, it can be concluded that the reflection and transmission capabilities of the turbine are not being affected by the temperature conditions of the flow.

- It is also relevant the significant difference of performance when operating in different MFR. In this sense, it is quite evident that the turbine is more permeable for transmitting the incident pressure waves to the outlet when operating at MFR 0.5, while the reflection is higher in general when the flow is not equally distributed in the inlets (MFR 0.3 and 0.65 approx.).

- Regarding the transmission coefficient result, there is an important difference between the performance of the shroud and hub turbine inlets (blue and red). A more concentration of points in the lower part of the graph can be observed for the shroud entry results (pulsating in shroud), while higher transmission coefficients are present when pulsating in hub.

Even though the MFR is a good variable to analyse the turbine performance, it might not be the most illustrative when looking for trends in the transmission and reflection losses and, thus, some other variables can be analysed. In Figure 3.16 the transmitted and reflected coefficients are correlated with the turbocharger speed, whereas in Figure 3.17 the same coefficients are confronted to the mass flow.

From Figure 3.16 it can be inferred that it appears to be an inversely proportional relationship between the reflection and transmission losses that correlates with the turbocharger speed. From the initial 90 krpm until approximately 130 krpm, the trend indicates a reduction in the transmitted coefficient and an increase in the reflected coefficient when the turbocharger speed increases. This trend apparently vanishes when the speed starts to get over the 130 krpm level, which might be an indication that some critical flow speeds are reached. In virtue of this observations, it can be stated that there is a trade-off between reflection and transmission losses that has to be dealt with when selecting the range of operation of the turbocharger in the calibration stage.

Results shown in Figure 3.17 can complement the previously observations about the turbocharger speed and the trade-off between reflection and transmission losses. Indeed, taking for example the pulsating-in-hub cases, the trend shows again a inversely proportional relation between the reflected and transmitted coefficients. This way, when the mass flow increases, the reflected coefficient also increases, whereas, on the contrary, the transmitted coefficient decreases.

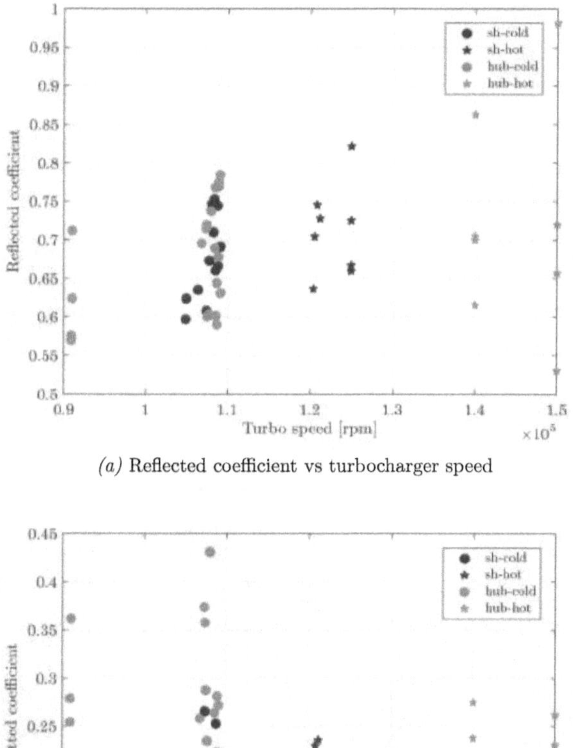

(a) Reflected coefficient vs turbocharger speed

(b) Transmitted coefficient vs turbocharger speed

Figure 3.16: Summary of all cases compared to turbocharger speed: reflected coefficient (a), transmitted coefficient (b).

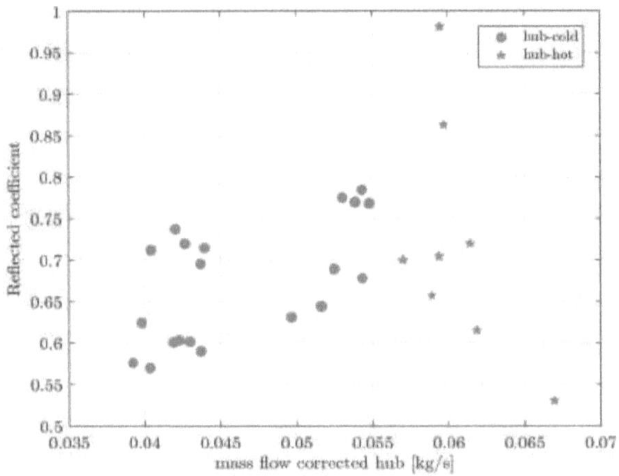

(a) Reflected coefficient vs mass flow

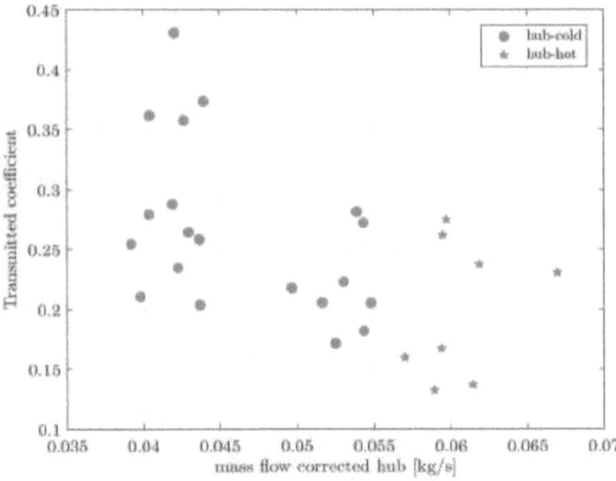

(b) Transmitted coefficient vs mass flow

Figure 3.17: Summary of hub cases compared to mass flow rate: reflected coefficient (a), transmitted coefficient (b).

After a general view of all the operating points, it might be interesting to analyse the differences that appear when comparing the same operating point but with the pulses performed in different branches. To this manner, in each of the following graphs, the result of a same operating pulsating point but with the pulses performed in shroud or hub entries is presented , again, blue and red respectively. The fact that the same pulse is imposed but in a different branch will allow to analyse differences in the response that are originated mainly due to the geometry of the entries and volutes of the turbine. In Figure 3.18, Figure 3.19 and Figure 3.20, the reflected, transmitted and indirect transmitted pressure components are shown using time as the independent variable. The time domain is used in order to look for particularities and differences in the pulse in terms of shape, amplitude, pulse-averaged range, etc.

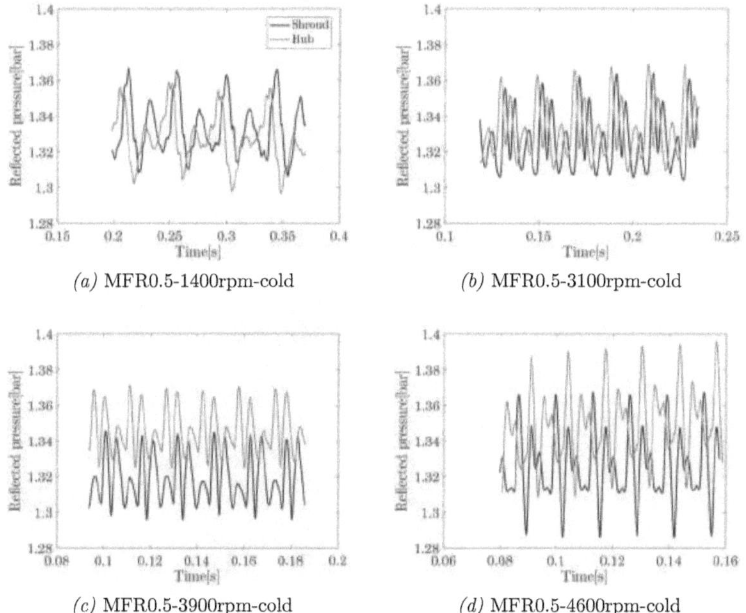

(a) MFR0.5-1400rpm-cold   (b) MFR0.5-3100rpm-cold

(c) MFR0.5-3900rpm-cold   (d) MFR0.5-4600rpm-cold

Figure 3.18: Reflected component in time domain for cases #1 and #17 (a), cases #4 and #21 (b), cases #7 and #25 (c), and cases #10 and #29 (d).

According to the results of the comparison between branches in the time domain, it can be stated that the oscillations are very similar in both shape and amplitude for incident and transmitted (Figure 3.18 and Figure 3.19). However, the indirect transmitted component (Figure 3.20) seems to present bigger differences depending on whether the pulse is generated in the shroud or hub branch. The conclusions are the same for all the four pair of cases analysed. As it can be observed, although the amplitude and the shape are fairly similar, the pulse-averaged pressure value that is transmitted to the shroud when pulsating in the hub is 20-40 mbar higher than the component that is indirectly transmitted to the hub when the pulses are performed in the shroud. This indicates that, with the geometry of this specific turbine, assuming that the same pulse is imposed as input in the entry, the indirect transmitted pressure that will travel back through the passive branch all the way back to the engine, will be higher in amplitude when pulsating in the hub branch. This kind of information might help taking better decisions in order to calibrate turbine operating range in terms of the mass flow distribution for the unequal admission operating points during the engine design stage.

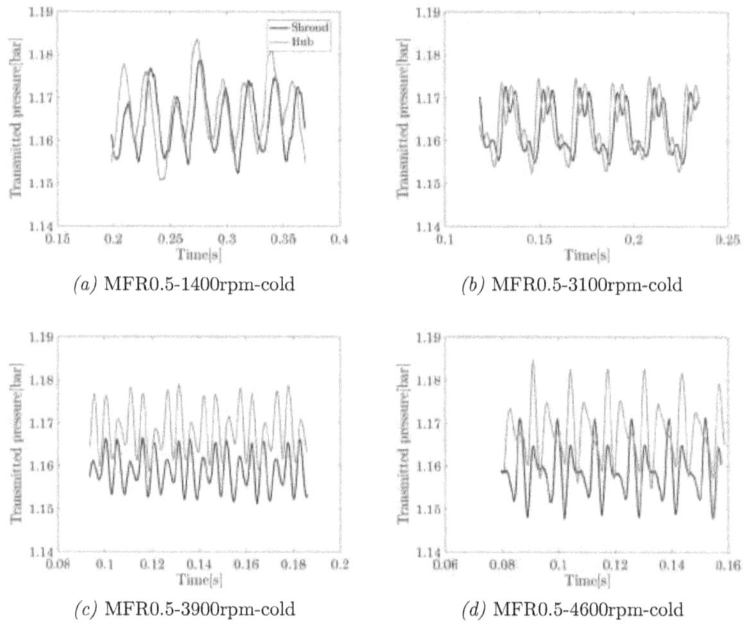

(a) MFR0.5-1400rpm-cold   (b) MFR0.5-3100rpm-cold

(c) MFR0.5-3900rpm-cold   (d) MFR0.5-4600rpm-cold

Figure 3.19: Transmitted component in time domain for cases #1 and #17 (a), cases #4 and #21 (b), cases #7 and #25 (c), and cases #10 and #29 (d).

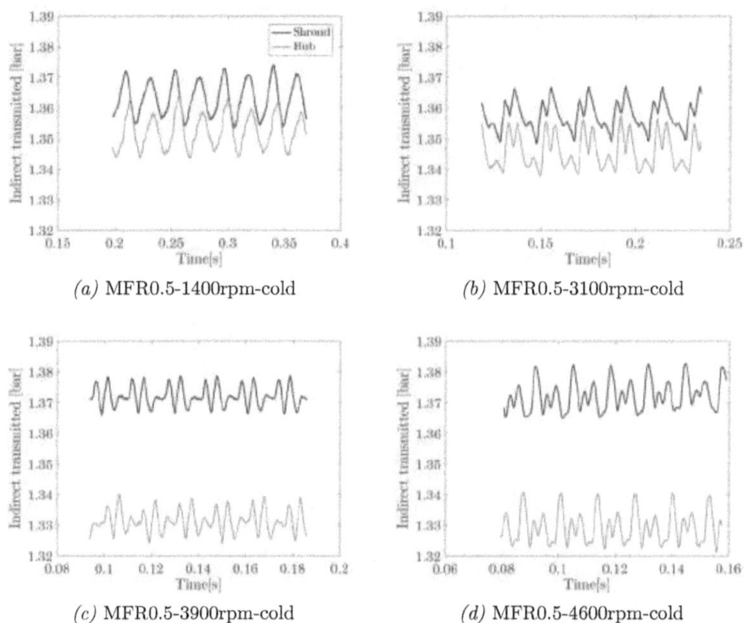

(a) MFR0.5-1400rpm-cold   *(b)* MFR0.5-3100rpm-cold

(c) MFR0.5-3900rpm-cold   *(d)* MFR0.5-4600rpm-cold

Figure 3.20: Indirect transmitted component in time domain for cases #1 and #17 (a), cases #4 and #21 (b), cases #7 and #25 (c), and cases #10 and #29 (d).

Continuing with the comparison between shroud and hub branches for the same pairs of operating points, it is also interesting to study an example of how evident is this difference of behaviour between branches along the frequency domain. in Figure 3.21 the evolution of the transmission loss in the frequency domain is presented for the full admission operating points. The transmission loss is defined as the difference between the incident and the transmitted amplitudes in dB.

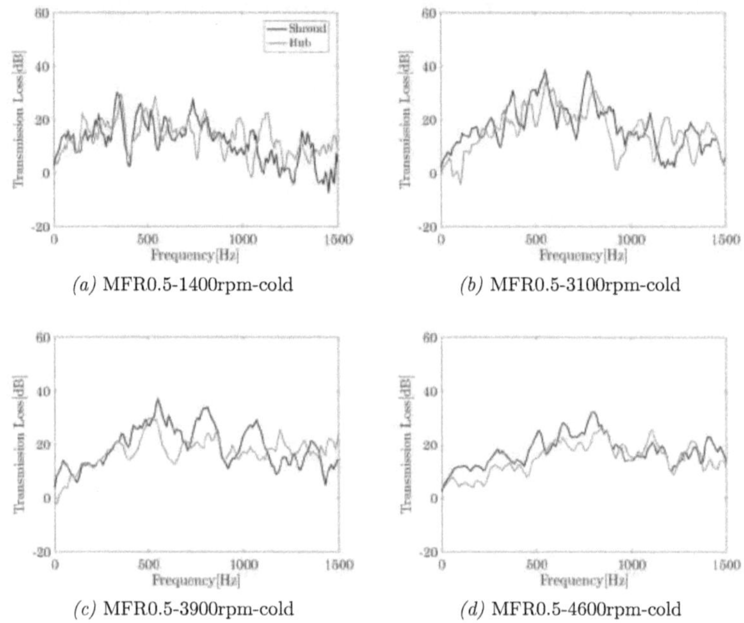

(a) MFR0.5-1400rpm-cold   *(b)* MFR0.5-3100rpm-cold

(c) MFR0.5-3900rpm-cold   *(d)* MFR0.5-4600rpm-cold

Figure 3.21: Transmission Loss in frequency domain for cases #1 and #17 (a), cases #4 and #21 (b), cases #7 and #25 (c), and cases #10 and #29 (d).

From the results of Figure 3.21, it might be stated that, in the majority of the cases, the transmission loss through the shroud branch is higher than in the hub branch with maybe the exception of the case in 3.21(a), where the low equivalent engine speed (1400 rpm) and the low amplitude might have affected the experimental accuracy. This tendency appears to be coherent with Figure 3.14, where it can be observed how the points corresponding to pulse-generation in shroud are in the lower range of transmission coefficient than the points where the pulse-generation is in hub for all the mass flow ratios.

3.4 Conclusions

The specifically built test-rig and a methodology of contrasted robustness [1], result in an innovate technique for testing two-scroll turbines. In spite of the challenge that unsteady flow always implies, the quality and accuracy of the results is very satisfactory, adding the novelty of an acoustic analysis for highly pulsating hot conditions with non-linear response. Furthermore, the flexibility of the gas stand allows to change the flow operating conditions with a controlled and precise procedure as well as to switch the pulsating branch easily.

The beamforming technique has been used to obtain the pressure wave decomposition through both turbine inlet branches and turbine outlet. Especially significant is the effect that the pulsating inlet branch produces in the other turbine inlet. The transmitted and "indirectly" transmitted components mainly affect the pumping losses as well as the total fuel consumption and the silencer and aftertreatment designs. The acquisition of realistic high quality data is nowadays crucial for the fitting and validation of the increasingly important numerical models.

From the point of view of the results of the turbine tested during the experimental campaign, the mass flow ratio seems to be a correct independent variable for analysing the evolution of the transmission and reflection parameters. The pressure signals seem to be transmitted to a higher degree when the MFR approaches 0.5. Although more research should be performed in order to fully explain this phenomenon, it might be related to a lower pressure loss between the volute and the rotor in the full admission case as the flow is more adapted at the rotor inlet. It is also worth noting the high values of the indirect transmission coefficient: in some cases, the indirect transmission pressure pulse has an amplitude that is almost 0.3 times that of the incident pressure pulse. Although the twin volute configuration helps to isolate different cylinders during their exhaust processes, it is clear that the interference effects might not be totally negligible.

The comparison between the non-linear acoustic responses of each branch can also add valuable information. The different location of each volute in the housing as well as the path and orientation followed by the flow when it hits the rotor causes differences in the performance. This information could be useful when configuring the turbocharger matching, as the low frequency oscillations travelling upstream of the turbine through the engine will affect the engine volumetric efficiency and even the turbocharger operating point. In addition, a better prediction of the high frequency oscillations in the exhaust line could result in a more efficient design of the muffler. Finally, experimental results under unsteady conditions will help calibrating and validating the models.

Chapter 3 References

[1]   J. Galindo, F. J. Arnau, L. M. García-Cuevas, and P. Soler. "Experimental va-lidation of a quasi-two-dimensional radial turbine model". *International Journal of Engine Research* (2018). ISSN: 1468-0874. DOI: 10.1177/1468087418788502 (cit. on pp. xi, 8, 57, 62, 84, 90, 113, 131, 133, 139, 177).

[6]　J. Serrano. "Imagining the Future of the Internal Combustion Engine for Ground Transport in the Current Context". *Applied Sciences* 7(10) (2017), p. 1001. ISSN: 2076-3417. DOI: 10.3390/app7101001 (cit. on pp. 3, 56).

[37]　A. Romagnoli, C. D. Copeland, R. Martinez-Botas, M. Seiler, S. Rajoo, and A. Costall. "Comparison Between the Steady Performance of Double-Entry and Twin-Entry Turbocharger Turbines". *Journal of Turbomachinery* 135(1) (2013). ISSN: 0889-504X. DOI: 10.1115/1.4006566 (cit. on pp. 8, 56).

[41]　M. S. Chiong, M. A. Abas, F. X. Tan, S. Rajoo, R. Martinez-Botas, Y. Fujita, T. Yokoyama, S. Ibaraki, and M. Ebisu. "Steady-State, Transient and WLTC Drive-Cycle Experimental Performance Comparison between Single-Scroll and Twin-Scroll Turbocharger Turbine". In: *WCX SAE World Congress Experience.* SAE International, 2019. DOI: https://doi.org/10.4271/2019-01-0327 (cit. on pp. 8, 56).

[42]　S. Rajoo, A. Romagnoli, and R. F. Martinez-Botas. "Unsteady performance analysis of a twin-entry variable geometry turbocharger turbine". *Energy* 38(1) (2012), pp. 176–189. ISSN: 03605442. DOI: 10.1016/j.energy.2011.12.017 (cit. on pp. 8, 57, 91).

[79]　J. Galindo, P. Fajardo, R. Navarro, and L. M. García-Cuevas. "Characterization of a radial turbocharger turbine in pulsating flow by means of CFD and its application to engine modeling". *Applied Energy* 103(0) (2013), pp. 116 –127. ISSN: 0306-2619. DOI: 10.1016/j.apenergy.2012.09.013 (cit. on pp. 21, 56, 90, 130).

[85]　S. Rajoo and M.-B. Ricardo. "Variable Geometry Mixed Flow Turbine for Turbochargers: An Experimental Study". *International Journal of Fluid Machinery and Systems* 1(1) (Oct. 2008), pp. 155–168. DOI: 10.5293/IJFMS.2008.1.1.155 (cit. on pp. 21, 57).

[86]　K. G. Hohenberg, P. J. Newton, R. F. Martinez-Botas, M. Halamek, K. Maeda, and J. Bouilly. "Development and Experimental Validation of a Low Order Turbine Model Under Highly Pulsating Flow". In: *Turbo Expo: Power for Land, Sea, and Air, Volume 2C: Turbomachinery.* ASME. June 2017, V02CT44A023. DOI: 10.1115/GT2017-63983 (cit. on pp. 21, 57, 90, 130).

[87]　J. R. Serrano, A. Tiseira, L. M. García-Cuevas, L. Inhestern, and H. Tartoussi. "Radial turbine performance measurement under extreme off-design conditions". *Energy* 125 (2017), pp. 72–84. DOI: 10.1016/j.energy.2017.02.118 (cit. on pp. 21, 57, 90, 135).

[88]　A. J. Torregrosa, A. Broatch, R. Navarro, and J. García-Tíscar. "Acoustic characterization of automotive turbocompressors". *International Journal of Engine Research* 16(1) (2015), pp. 31–37. DOI: 10.1177/1468087414562866 (cit. on pp. 21, 57, 90, 130).

[101] G Pinero, L Vergara, J. M. Desantes, and A Broatch. "Estimation of velocity
       fluctuation in internal combustion engine exhaust systems through beamforming
       techniques". *Measurement Science and Technology* 11(11) (2000), p. 1585. DOI:
       10.1088/0957-0233/11/11/307 (cit. on pp. 36, 39, 64, 133, 177).

[104] R. Hogg. "Life beyond euro VI". *automotiveworld* (2014). URL: http://www.
       automotiveworld.com/megatrends-articles/life-beyond-euro-vi/ (cit. on
       pp. 56, 90).

[105] M. Knopf. "How low can we go? Downsizing the internal combustion engine".
       Ingenia (2011). URL: https://www.ingenia.org.uk/Ingenia/Articles/
       11612c67-32db-477a-af8f-38cef44ee7b2 (cit. on pp. 56, 90).

[106] E. Watel, A. Pagot, P. Pacaud, and J.-C. Schmitt. "Matching and Evaluating
       Methods for Euro 6 and Efficient Two-stage Turbocharging Diesel Engine". In:
       SAE 2010 World Congress and Exhibition. SAE international. 2010. DOI: 10.
       4271/2010-01-1229 (cit. on p. 56).

[107] F. Pischinger and A. Wuensche. "The characteristic behaviour of radial turbines
       and its influence on the turbocharging process". In: *Proceedings of the CIMAC
       Conference, Tokyo, Japan*. 1977, pp. 545–568 (cit. on p. 56).

[108] A. Torregrosa, J. Serrano, J. Dopazo, and S. Soltani. "Experiments on wave trans-
       mission and reflection by turbochargers in engine operating conditions". *SAE
       Technical Papers* 2006(01:0022) (2006). DOI: 10.4271/2006-01-0022 (cit. on
       pp. 57, 90).

[109] A. Costall, V. Cheong, H. Flora, A. Munasinghe, R. Ivanov, R. W. Kruiswyk,
       and J. R. McDonald. "Development of a Novel Transient-Pulsating Flow Rig for
       Engine Air System Research using GT-SUITE". In: *European GT Conference*.
       2018 (cit. on pp. 57, 180).

[110] C. Arcoumanis, I. Hakeem, L. Khezzar, R. F. Martinez-Botas, and N. C. Bai-
       nes. "Performance of a Mixed Flow Turbocharger Turbine Under Pulsating Flow
       Conditions". In: *Volume 2: Aircraft Engine; Marine; Microturbines and Small
       Turbomachinery*. ASME, 1995, V002T04A011. ISBN: 978-0-7918-7879-8. DOI:
       10.1115/95-GT-210 (cit. on p. 57).

[111] C. Copeland, R. Martinez-Botas, and M. Seiler. "Unsteady performance of a
       double entry turbocharger turbine with a comparison to steady flow conditions".
       In: *Proceedings of the ASME Turbo Expo*. Proceedings of the ASME Turbo Expo.
       ASME, 2008. DOI: 10.1115/GT2008-508279 (cit. on p. 57).

[112] M. Yang, K. Deng, R. Martines-Botas, and W. Zhuge. "An investigation on
       unsteadiness of a mixed-flow turbine under pulsating conditions". *Energy Con-
       version and Management* 110 (2016), pp. 51–58. ISSN: 01968904. DOI: 10.1016/
       j.enconman.2015.12.007 (cit. on p. 57).

[113] C. D. Copeland, R. F. Martínez-Botas, and M Seiler. "Comparison between steady and unsteady double-entry turbine performance using the quasi-steady assumption". *Journal of Turbomachinery: Transactions of the ASME* 133 (3 2010). DOI: 10.1115/1.4000580 (cit. on p. 57).

[114] M. Yang, R. F. Martinez-Botas, S. Rajoo, T. Yokoyama, and S. Ibaraki. "Influence of Volute Cross-Sectional Shape of a Nozzleless Turbocharger Turbine Under Pulsating Flow Conditions". In: *Volume 2D: Turbomachinery*. ASME, 2014, V02DT42A025. ISBN: 978-0-7918-4563-9. DOI: 10.1115/GT2014-26150 (cit. on pp. 57, 91).

[115] M. Cerdoun and A. Ghenaiet. "Unsteady behaviour of a twin entry radial turbine under engine like inlet flow conditions". *Applied Thermal Engineering* 130 (2018), pp. 93–111. ISSN: 1359-4311. DOI: 10.1016/j.applthermaleng.2017.11.001 (cit. on pp. 57, 91).

[116] C. Cravero, D. De Domenico, and A. Ottonello. "Numerical Simulation of the Performance of a Twin Scroll Radial Turbine at Different Operating Conditions". *International Journal of Rotating Machinery* 5302145 (2019). ISSN: 15423034. DOI: 10.1155/2019/5302145 (cit. on p. 57).

[117] R. Kabral, Y. El Nemr, C. Ludwig, R. Mirlach, P. Koutsovasilis, A. Masrane, and M. Åbom. "Experimental acoustic characterization of automotive twin-scroll turbine". In: *12th European Conference on Turbomachinery Fluid Dynamics and Thermodynamics ETC*. 2017, p. 363 (cit. on pp. 57, 75, 92).

Chapter 4

One dimensional twin-entry radial turbine model for non-linear pulse simulations

Contents

4.1 Introduction

The increasing stringent pollutant emissions legislations and the demand of low consumption and low emission vehicles has lead the automotive industry towards highly efficient engines. During the past few years, the turbocharging technology together with the downsized and rightsizing approach has been proved to be hugely effective, reducing dramatically the CO_2 emissions among other pollutants [105]. The high percentage of recoverable energy available in the exhaust manifold [104] can be used in the turbine for driving the compressor and boosting the intake air for a more efficient combustion.

In spite of all the advantages of the use of turbochargers, the exhaust flow energy can be insufficient in some low engine speed operating conditions, leading to a lack of engine performance during transients. With its particular design of the housing, in which the flow is divided into two separated flow channels fed by separate exhaust banks, the two-scroll turbine decouples the energy pulses of alternative firing order cylinders, using the dynamic pressure effectively for optimising the scavenging effects. In this manner, multiple scroll turbines allow a better recovery of the dynamic pressure of the gas pulses [118] as they can avoid part of the backflows generated by interferences of exhaust processes between consecutive firing order cylinders and, thus, improve the engine volumetric efficiency. Due to their big valves overlap, important interferences can take place even in 4 cylinder engines. But not only gasoline engines, big diesel engines for truck and long haul engines with more than 4 cylinders, typically 6 cylinders, suffer from significant overlapping effects and improve significantly their transient response when being turbocharged using two-scroll turbines. Fast one-dimensional, time-resolved simulations of two-scroll turbines under pulsating flow conditions become a valuable tool to analyse their behaviour during engine design and optimisation. In this chapter, a model thought for this kind of application is described and experimentally validated with the tests obtained in chapter 3.

4.1.1 Literature review

Due to the opening and closing of the cylinder valves during real driving conditions, the turbine is constantly operating under unsteady conditions [79]. And, thus, it is fundamental to correctly comprehend its behaviour not only in steady state but also in pulsating conditions. With this purpose, Winterbone et al. [119] and Winterbone and Pearson [120] performed a rigorous review on the topic. Investigators have also performed different experiments for pulsating flow in single entry turbines, such as Torregrosa et al. [88] or Serrano et al. [87] for unsteady and off-design performance, Torregrosa et al. [108] for diesel engines or Hohenberg et al. [86]. Piscaglia et al.[121] show an interesting approach, measuring the instantaneous pressure, temperature and mass flow in different points in the turbine inlet duct and the turbine outlet duct. Galindo et al. [1] present a similar experimental methodology as the one here presented, being applied for a variable geometry turbine. Finally, Cappelaere et al. [122] present a new test methodology for testing turbochargers under highly pulsating conditions.

Pulsating flow creates, however, unique operating characteristics in a two-scroll turbine compared to a single entry turbine that, furthermore, are highly sensitive to the flow distribution through the turbine entries. Baines et al. [123] studied the performance of a mixed flow vaneless radial turbine under pulsating conditions, performing pulses of different frequencies (20 – 60 Hz), and turbine speeds (300 – 500 rps). A more recent improved version of the experiments presented by Baines et al. [123] can be found in the work by Rajoo et al. [42], in which they were able to characterise the turbine behaviour under a given pulse flow, analysing also different vane angle settings as it was also exposed in previous works [43]. Some other contributions to the analysis of the twin and double entry performance under pulses are the ones carried out by Copeland et al. [124, 45], where, in addition to the unsteady performance, a comparison with the turbine steady state performance is also presented. Costall et al. [36] documented some similar findings including the unsteady behaviour of a twin entry turbine for heavy duty non-road applications in full and unequal admission conditions.

As it has been noticed by many researchers, it is considerably challenging to obtain reliable experimental results when measuring real pulse engine conditions. Because of that, some interesting studies have been performed using CFD that have contributed to the understanding of the turbine unsteady phenomena. M. Yang et al. [114, 125] presented the investigation of the influence of volute cross-sectional shape on the performance of a radial turbocharger turbine under pulsating conditions supported with experimental results. Copeland et al. [126] presented an investigation on the effect of the unequal admission on the performance of a double-entry turbine by means of a full 3D computational fluid dynamics (CFD) model. Most recently, M. Cerdoun and A. Ghenaiet [115] performed a study that characterises the flow behaviour of a twin-entry turbine under engine-like inlet flow conditions with a focus on the development of Dean vortices and the hysteresis phenomena.

CFD modelling application is, however, limited by its huge computational cost. In contrast, 1D and quasi-1D models provide considerable accuracy with a much more manageable time required, reaching even real time capabilities for some applications. One-dimensional gas dynamic modelling includes a vast range of variety and complexity among the different approaches when analysing the pulse performance. A first and computationally very affordable possibility is the "filling-emptying" model, as the ones presented by Baines et al. [123] and Payri et al. [23] In these cases, the volume of the twin-scroll housing is considered, what enables to predict the mass and energy accumulation effects. Nevertheless, in the work presented in this chapter, the focus is set in the turbine acoustic characterisation with non-linear capabilities and, thus, a wave action pulse flow model has been used. With this approach, and being furthermore application oriented, the works by Costall et al. [36] and Chiong et al. [38] can be found. In 2015, a more complex model was presented [127], using a loss parameters tuning criteria for twin-scroll modelling based on an approximation of the real physics. Following previous works developed in Imperial College, M. Chiong et al. [39] explore the potential improvement in the twin-entry pulse flow modelling by considering the unequal and partial admission conditions. Finally, Chiong et al. [128] shows a revised 1D pulse flow modelling method to predict the twin-scroll turbine performance that is validated against experimental data, providing the technique to generate the full range of admission curves from the full admission experimental data.

Especially interesting is the contribution made by Kabral et al. [117], where an experimental methodology for performing an acoustic characterisation of the turbine with a pressure decomposition method is described, and the linear amplitude dynamic response of the twin-scroll turbine is obtained.

The technology described in detail in chapter 3 provides the experimental method to analyse the non-linear response under different conditions (hot and cold), with different amplitudes and equivalent engine points of operation separating the effects of each branch independently. In this chapter, a one-dimensional quasi-steady model is validated against the pulsating experimental data obtained in chapter 3, providing a very good degree of correlation, that allows to predict accurately the characteristics of the pulse transmitted to the exhaust line and also reflected to the exhaust manifold, which implies more efficient decisions during the design and matching stages between turbocharger and engine.

The pressure waves prediction in the frequency domain keeps good accuracy for up to 1000 Hz or even more, while keeping the low computational cost. In single threaded simulations, the model needs 60 seconds for each single second of simulated time in an x86 CPU running at 3.2 GHz. It can easily be used coupled with other submodels of the turbocharger, taking into account the compressor quasi-steady map extrapolation [31], the heat transfer [27, 28, 29] or the mechanical losses in the bearings [32].

The model exposed in this chapter has been developed in coordination with other works about modelling two-scroll turbines carried out in CMT-Motores Térmicos. More specifically, the steady-state experiments presented in [129] and the turbine map extrapolation and interpolation model documented in [34] have been used as a base in order to develop the work that is shown in this chapter.

4.1.2   Experimental campaign

The experimental facility is aimed to perform turbine acoustic characterizations under non-linear pulse conditions by means of a rotating valve for generating the pulses and an array of instantaneous pressure sensors (piezoelectric) for the pressure decomposition. The method that has been followed in order to obtain the data used to validate this model was described in detail in chapter 3. Nevertheless, a basic sketch of the gas stand is again presented in Figure 4.1 for clarity's sake. Summarizing the experimental procedure, a two-stage radial compressor is used to provide the air mass flow to the

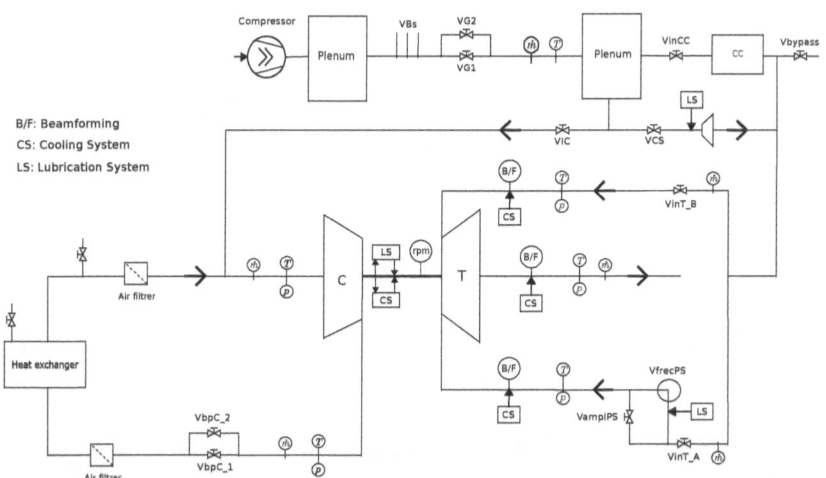

Figure 4.1: Turbocharger gas stand facility layout at CMT–Motores Térmicos.

turbine, and the high volume plenums are aimed to stabilise the flow. When operating
in hot conditions, a combustion chamber can easily be activated to heat the air to the
desired turbine inlet temperature. For generating pulses of high non-linear amplitudes,
an especially designed rotating valve is situated in one of the inlet branches and a plate
with orifices is used for obtaining the desired pulse frequency. Although the rotating
valve is fixed in one of the turbine inlet branches, the flexible pipes installed before the
rotating valve allow to easily switch between inlet pulsating branches.

As previously mention in subsection 3.2.2, the turbine used in this study is a twin-
entry radial inflow turbine, with wastegate and asymmetrical entry that is designed to be
used in a D-segment passenger car, 2.0 L gasoline engine. Regarding the nomenclature
adopted, shroud and hub will differentiate each entry referring to its location inside
the housing. All the measurements of diameters and lengths were taken for analysis
and modelling purposes. For the volume of the volutes, a silicone mould was used
(Figure 4.2). The moulds and also the rotor have been scanned by means of a 3D
scanner and are currently being used for CFD studies that will be part of a future
book.

As it was previously indicated, the mass flow ratio, MFR, is defined as in Equation 4.1. In Table 4.1 the operating conditions used during the experimental campaign

Case label	Pulsating branch	Approximate MFR	Equivalent engine speed	Conditions
#1–#3	Shroud	0.5–0.7	1400 rpm	Cold
#4–#6	Shroud	0.5–0.7	3100 rpm	Cold
#7–#9	Shroud	0.5–0.7	3900 rpm	Cold
#10–#12	Shroud	0.5–0.7	4600 rpm	Cold
#13–#16	Hub	0.5	1400 rpm to 4600 rpm	Cold
#17–#20	Hub	0.5	1400 rpm	Cold
#21–#24	Hub	0.3–0.5	3100 rpm	Cold
#25–#28	Hub	0.3–0.5	3900 rpm	Cold
#29–#32	Hub	0.3–0.5	4600 rpm	Cold
#33–#40	Shroud	0.7	1900 rpm to 5600 rpm	Hot
#41–#48	Hub	0.3	1900 rpm to 5600 rpm	Hot

Table 4.1: Detailed list of the experimental points for two-scroll turbine experiments.

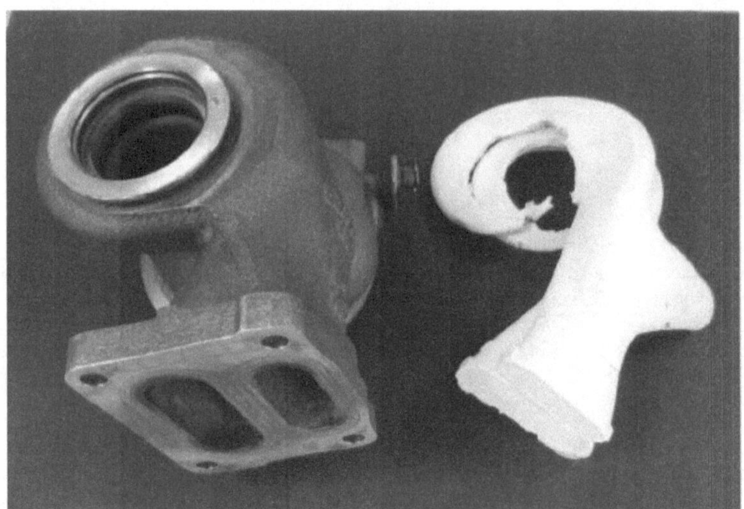

Figure 4.2: Turbocharger specimen and silicone mould.

are summarised.

$$MFR = \frac{\dot{m}_{\text{shroud}}}{\dot{m}_{\text{shroud}} + \dot{m}_{\text{hub}}} \tag{4.1}$$

4.1.3 Quasi-adiabatic and non-linearity assessment

The experiments were performed in quasi-adiabatic conditions, minimising the external and internal heat transfer in the turbocharger and in the inlet and outlet ducts. Also, several thermocouples were used to get the average temperatures, thus minimising the typical inaccuracies found when measuring the turbine outlet temperature[130, 131]. The overall energy balance was used to check the quality of the results, as explained by Serrano et al. [129].

In Figure 4.3 the relation between the power of the external heat and the turbine power is shown, providing, directly, the expected error in the calculation of the turbine power output due to the external heat flow losses. The heat flow to the ambient is situated below 0.5 % of the measured turbine power output, which indicates an effective thermally insulation of the ducts in the controlled laboratory conditions. Thus, the heat flow to the ambient is low enough to consider quasi-adiabatic experimental conditions for all the operating points.

To estimate the heat flow to the ambient, Equation 4.2 is used:

$$\dot{Q}_{\mathrm{env}} = A_{\mathrm{surf}} \cdot h_{\mathrm{conv}} \cdot (T_{\mathrm{surf}} - T_{\mathrm{env}}) \tag{4.2}$$

where T_{env} is the environment temperature, T_{surf} is the maximum surface temperature of the insulation, A_{surf} is the exposed surface of the insulation, and the convective heat transfer coefficient h_{conv} is $10\,\mathrm{W\,m^{-2}\,K^{-1}}$. Note that the temperature difference between the insulation and the surroundings was low enough that the radiation was much smaller than the external convection.

On the other hand, the turbine power is estimated using the total enthalpy flow as in Equation 4.3:

$$\dot{W} = \dot{m} \cdot c_{\mathrm{p}} \cdot (T_{\mathrm{6t}} - T_{\mathrm{0t}}) - \dot{Q}_{\mathrm{int}} \tag{4.3}$$

where $\dot{Q}_{\mathrm{int}}$ expresses the internal heat losses from the gas to the turbine, calculated as explained by Serrano et al. [67].

It is worth noting that the experimental data used to validate this model has been obtained in a gas stand using pressure pulses that have been generated by means of a rotating valve. It is then necessary to assess the amplitude and non-linearity of the pulses generated. For this purpose, chapter 3 gives a mathematical proof of the non-linearity of the excitations in the non-linearity assessment (subsection 3.3.2), indicating that the pulses were strong enough to induce realistic non-linear conditions in the turbine. The experimental data obtained is then reliable enough to evaluate the capability of the model presented in this work to predict the turbine behaviour under high-amplitude pulsating flow typical of engine operation.

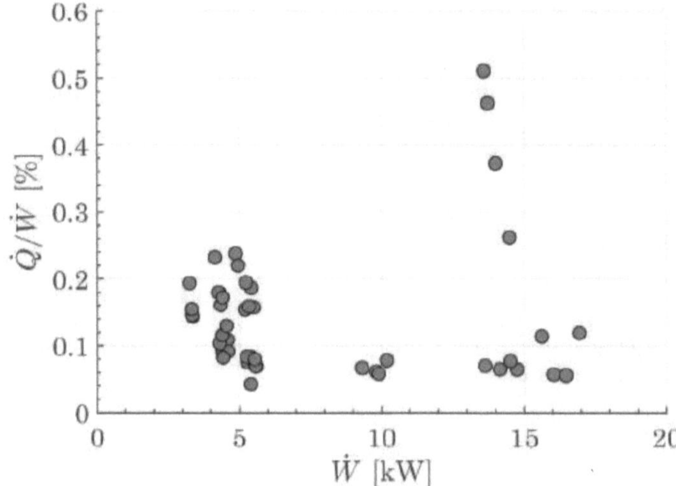

Figure 4.3: Relative external heat flow.

4.2 Model description and quasi-steady adaptation

The model section of this chapter is organised as follows. Firstly, the model domain utilised and the method of calculation is summarised. Then, the twin-turbine 1D model developed in CMT is presented. Finally, the modelling adaptations that have been necessary to provide the model with unsteady capabilities are detailed. The work presented in this book is focused in the turbine performance under pulsating conditions, because of that, the modelling and validation work exposed in this chapter is mainly centred in pulsating flow.

4.2.1 Model domain

In order to build the turbine model domain, the previous experience in the 1D approach for single entry turbine modelling described by Serrano et al. [25] has been used. Also it has been used the previous experience in single entry turbine efficiency and mass flow parameters extrapolation [30, 132]. The contrasted robustness and accuracy of the previous single entry version of the model reinforced the idea of a simple two-volutes adaptation of the existing model. The main source of obstacles during the modelling process will emanate from the flow effects that the interaction between the two scrolls will produce.

In Figure 4.4 a basic sketch of the model domain is presented. Straight ducts have been used for the inlets and outlet pipes. The volutes are also one-dimensional, with an equivalent length of half of the developed volute, as first described originally by Chen and Winterbone [24, 133]. The stator is modelled as a zero-dimensional element with a volume equal to that of the physical stator and rotor combined, to take into account their mass and energy accumulation effects. For simulations where waste-gated turbines are used, this volume will consider the vaneless space between the wheel and the volute. The model shown in Figure 4.4 covers the computational flow domain from the measurement plane in the inlet (station 0), till the open end in the gas stand outlet pipe (station 7). Station 1 indicates the beginning of the turbine with the inlet duct, while station 2 references the volute inlet. Station 3 is the volute outlet and station 4 is the equivalent zero-dimensional element outlet. Station 5 represents the conditions after the rotor. Stations 6 and 7 are the turbine outlet and the domain outlet, respectively.

Using straight ducts to simulate the volute flow with an equivalent length equal to half of the developed volute keeps the complexity of the model low while reproducing with high accuracy one of the main sources of linear and non-linear acoustics effects of the turbine. Also, although using three equivalent nozzles and an intermediate volume instead of just two equivalent nozzles seems to introduce some complexity, it has several important advantages:

- It is closer to the real geometry of the turbine.

- It allows to model some mass and energy accumulation, wave and thermal transient effects.

- The effect that the addition of an intermediate volume produces has been previously discussed in chapter 2, producing significant improvements in the prediction for the middle and high frequency spectra.

- It reduces the stabilisation and initialisation issues that may arise during the calculations due to its damping effects.

- In the case of the need to speed up the calculation evolving this model to a 0D approach, this model gives quite a straight-forward solution.

In the following sections the model will be further detailed. The focus of this section is, however, the adaptation and modification of the baseline steady-state model in order to provide it with unsteady capabilities.

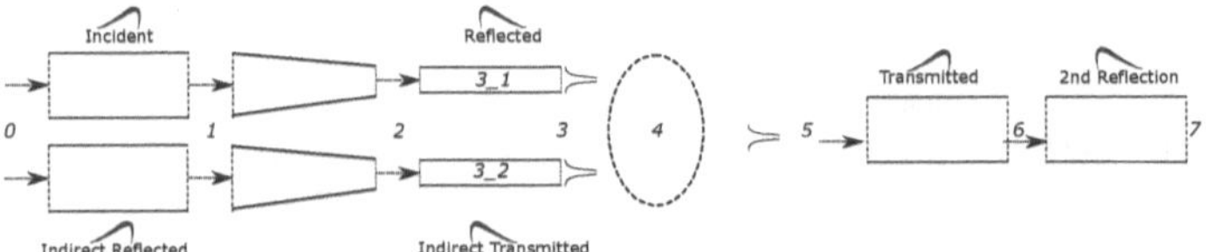

Figure 4.4: Twin turbine model computational domain. The main sections are labelled with numbers and referred to as *stations* in text.

4.2.2 Method of calculation

Following the procedure of the rest of the **book**, for this study the simulation results are obtained using OpenWAM [17], developed in CMT–Motores Térmicos. Again, the one-dimensional elements, such as the turbine inlets and outlet ducts, are discretised by means of a finite-volume approach and computed using a Godunov scheme. Please refer to section 2.2, where the Euler equations for fluid dynamics are exposed and the computational methods and solvers used are detailed.

4.2.3 Baseline model

The objective of this section is to provide an understandable reference framework for exposing the necessary modelling steps described in the next section, that have been followed in order to evolve the model to a quasi-steady integrated model with non-linear unsteady capabilities.

In this model approach, which is based on hypothesis presented by Serrano et al. [30], the turbine can be modelled as an equivalent nozzle with an equivalent area. This area varies on the operative conditions in a way that, assuming an adiabatic expansion, it is possible to obtain a full range of equivalent areas for each vane position in variable geometry turbines. This effective area model mainly depends on some main geometrical parameters that are easily measurable, the limited range of measured points that the manufacturer provides, and the corresponding fitting coefficients. With a similar procedure, the efficiency model [132] is based on the Euler equation of turbomachinery for radial gas turbines assuming some simplifications. Again, it uses the limited map provided by the manufacturer and some geometrical information, to provide a complete set of efficiency maps for each VGT configuration that are related also by fitting coefficients.

With a new and innovative approach, the previous turbine extrapolation model [132] [30] can be adapted for two-scroll turbine. Using an approach of considering the two entries as individual turbines, it is possible to obtain a set of maps of effective area (and, thus, reduced mass flow) for each of the entries that will depend on the mass flow ratio instead of on the VGT position [129]. Once the equivalent nozzle area is obtained, the reduced mass flow in both entries, shroud and hub, can be calculated assuming an isentropic expansion through a nozzle. On the other hand, the efficiencies are obtained considering the enthalpy drop generated by each individual turbine as a difference between each inlet and the common outlet. Thus, two different set of efficiency maps are obtained for each inlet, that are related to the effective area maps. This steady-state interpolation and extrapolation model is described in detail by Serrano et al. [34].

4.2.4 Pulsating flow model capabilities

In this section, the different stations depicted in Figure 4.4 will be referred when needed but without explicit reference to Figure 4.4 for improving section readability. During unsteady simulations, the different one-dimensional elements are computed solving Euler equations of one-dimensional fluid dynamics, and the intermediate volume is computed solving its mass and energy conservation equations. The flow in stations 3, 4 and 5 is computed using a technique [25] to split the pressure ratio in the turbine between stator and rotor representing nozzles (stations 3 and 5 respectively). Heat transfer effects are taken into account as an energy source term changing the flow temperature when passing between stations 3 and 5, and an extra energy sink term is introduced in the intermediate volume 4 equal to the power output of the turbine each time step. The extrapolated turbine map [34] is used to compute the stator (3) and rotor (5) nozzles using techniques [25]. The efficiency is quasi-steadily obtained [34].

The first obstacle to solve when building this model was the mass flow time-marching calculation. To obtain the areas of the nozzles at any time step, the mass flow ratio, the pressure ratio between the inlet of station 3 and the outlet of station 5 for each branch, the reduced rotational speed of each branch and the isentropic efficiency of each branch are needed [34]. When simulating only steady state conditions, this approach is accurate enough for generating correct mass flow results. In unsteady flow simulations this may not be the case, as the conditions at the beginning of each time step are used to compute this effective section in an explicit time integration scheme. Unsteady flow simulations may generate zero or reverse flow in station 3 for one of the volute branches instantaneously and, as a consequence, a mass flow ratio equal to 0 or 1. The next time step, the effective section of that nozzle is computed with this mass flow ratio, which leads to an effective section equal to 0. This way, the mass flow ratio is again 0 or 1, and the process is repeated until the simulation finishes. For confronting this difficulty, instead of computing the mass flow ratio with the conditions in stations 3 and 5, the mass flow per branch is obtained using station 1. This way, whenever the effective section of one of the nozzles is equal to 0, it can change to a bigger value in the following time steps when a pressure gradient produces again any amount of mass flow in station 1.

Using the mass flow ratio from station 1 to compute the effective section downstream can introduce some instabilities on its own. In reality, any change in the conditions in station 1 should take some time to reach stations 3 and 5, of the order of 1 ms for the geometry tested for this research. In fact, filtering the mass flow ratio with a first order delay of 1 ms produced optimal results for the studied turbine. The sensitivity of the results to this delay is small, so only the order of magnitude is actually needed.

The single entry turbine effective section model computes a single-nozzle equivalent section [25] and, after that, shares this between two equivalent nozzles with an intermediate volume [25]. The steady-state pressure of this volume is computed from the turbine efficiency, the rotor geometry and the degree of reaction to get the equivalent section of the two nozzles. In the case of a twin-entry turbine, the steady-state pressure of the intermediate volume is computed as a mass flow ratio-weighted average of the steady-state pressure obtained for each turbine branch as a single entry turbine. This way, the effective section of each stator nozzle is computed, as well as two effective sections for the rotor nozzle, one for each branch. The final rotor nozzle effective section is the sum of the result of each branch. When the pressure difference between branches is high enough, backflow between them can appear naturally before the rotor nozzle.

The efficiency of each branch is computed as a function of its blade speed ratio, reduced speed, mass flow ratio and effective section [34]. As the effective section is a function of the efficiency and the efficiency is a function of the effective section, the results are obtained in an iterative process.

With the efficiency and the effective sections, the mass flow of the three nozzles is computed and used as boundary conditions for the different one-dimensional and zero-dimensional elements connected to them. With the mass flow of each stator nozzle and its efficiency, the power output of each branch is computed and used as an energy sink term in the intermediate volume. The final turbine outlet temperature, after mixing the gas flows of each individual branch, might be higher than that produced for one of the branches and lower than that produced by the other one if they were acting as separate turbines. The process for computing one single time-step is summarised in Figure 4.5:

4.3 Validation of the model

The objective of this section is to present the performance and potential of the quasi-steady two-scroll turbine model coupled with zero- and one-dimensional elements in terms of non-linear pulses prediction capabilities. To do so, the experimental campaign from chapter 3 first presented by Serrano et al. [2] is used to obtain the validation results that are next presented. Firstly, a "stationary" validation is presented as a previous step using averaged values of the pulsating tests as an early verification of the implementation. Next, the performance of the model is demonstrated using a turbine characterisation by means of a pressure decomposition under non-linear pulsating excitations. Finally, the instantaneous variation of two important parameters, i.e., the efficiency and the mass flow ratio, is exposed with the intention of making evident how determinant the correct turbine unsteady prediction might be, when it comes to engine-like operating conditions.

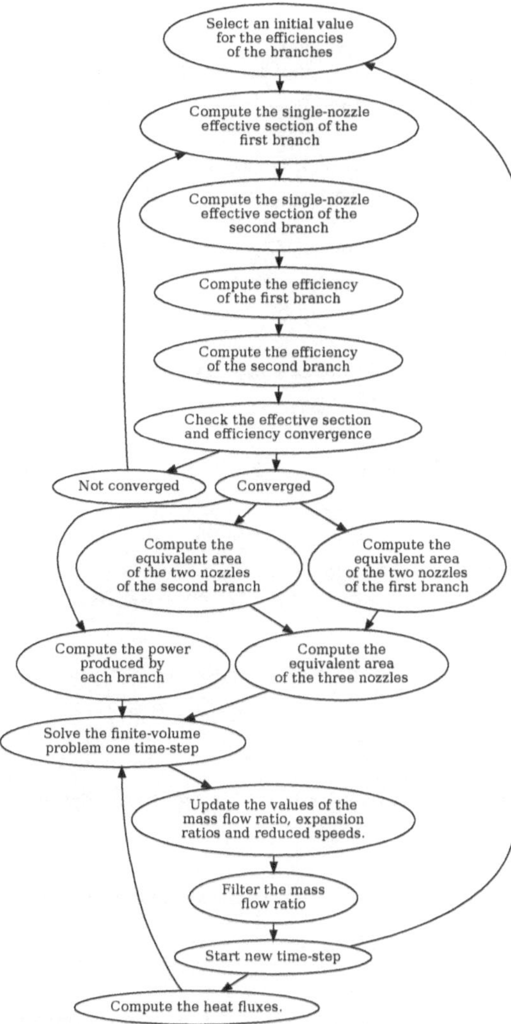

Figure 4.5: Summary of the steps needed to solve one time-step of the turbine model.

4.3.1 Steady-state validation

In addition to the experimental uncertainty and model accuracy, the fully-coupled model might introduce computation instabilities. Because of that, it was decided to start with an initial validation against averaged data from the pulsating flow campaign to assess the general accuracy and stability of the model.

Using the steady-state turbine map obtained as described by Serrano et al. [129], where quasi-adiabatic measurement are obtained at different mas flow ratios, the model was tested against averaged data. For performing this preliminary validation, the following inputs and outputs where established:

- Input: Averaged mass and total enthalpy flows are imposed at both turbine inlets.

- Input: Turbocharger speed is imposed.

- Input: Averaged static pressure at the turbine outlet is imposed.

- Output: Expansion ratio in both branches are compared against experimental data.

- Output: Turbine outlet temperature is compared against experimental data.

- Output: Turbine outlet mass flow rate is compared against experimental data.

For this particular validation only the cold conditions operating points have been used, as the aim was only to seek for procedure or implementation errors of the model with as simple as possible operating conditions. Hot operating cases will be properly validated and discussed in the following section.

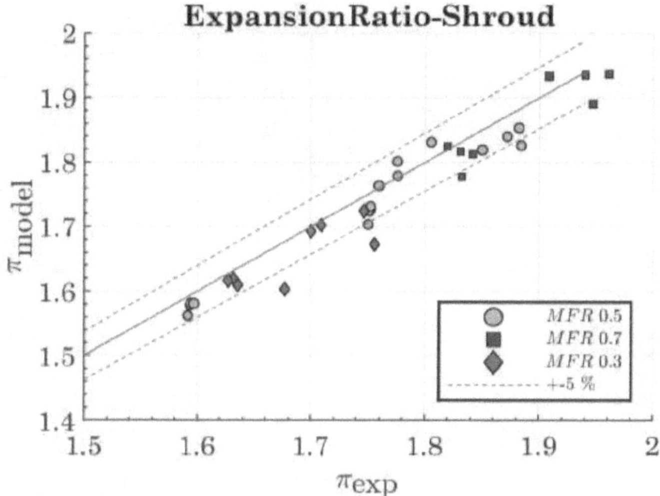

Figure 4.6: Correlation between experiment and model. Expansion ratio in shroud branch.

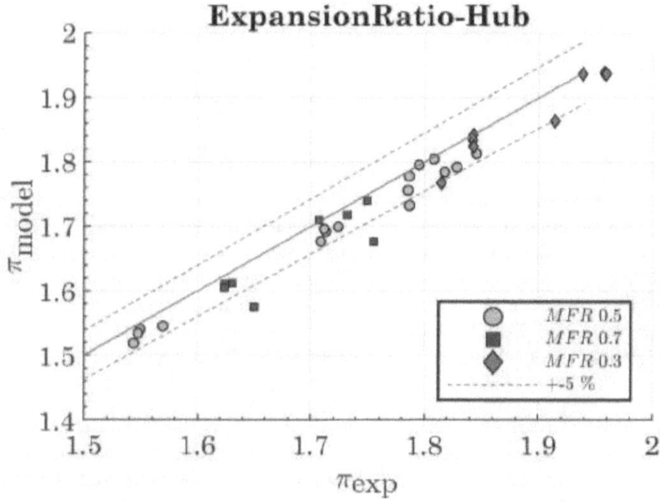

Figure 4.7: Correlation between experiment and model. Expansion ratio in hub branch.

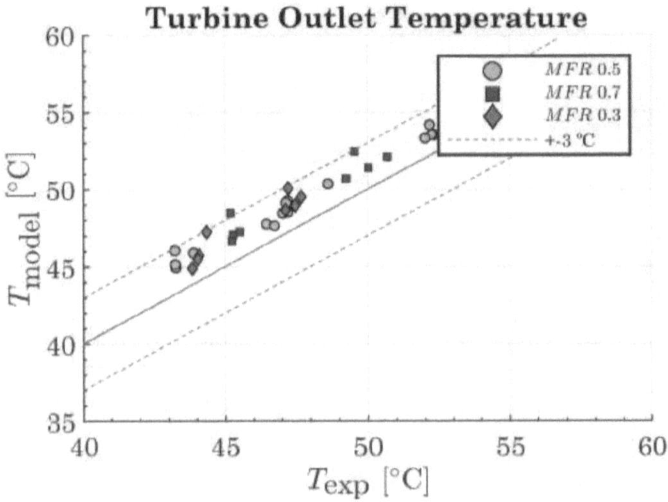

Figure 4.8: Correlation between experiment and model. Turbine outlet temperature.

To assure that the expansion in the turbine is well predicted in terms of pressure drop, the expansion ratio is evaluated separately in shroud and hub branches. In this sense, Figure 4.6 and Figure 4.7 show a good agreement between experiments and simulations, with a discrepancy below the 5 % level in the majority of tested conditions. As it can be observed in Figure 4.8, the turbine outlet temperature is also well predicted, with an error less than 3 K. Finally, Figure 4.9 also gives an error lower than 5 % for the outlet mass flow in this preliminary "stationary" results, with the majority of the points around the 2.5 % level of error.

The model did not present numerical instabilities. Some of the differences between the experimental data and the simulated results come from the averaging of the boundary conditions: the non-linear relationship between the expansion ratio and the mass flow of the turbine introduces discrepancies when comparing steady-state simulations with pulsating experimental data.

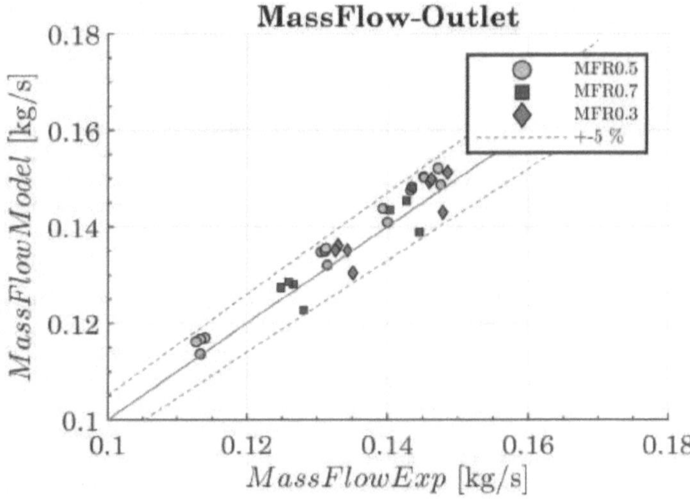

Figure 4.9: Correlation between experiment and model. Turbine outlet mass flow rate.

4.3.2 Pulsating validation

In the next paragraphs, two operating points obtained from the experimental campaign are presented as examples. These cases represent the performance of the model, and have been selected in a way that the operating conditions were diverse enough. This way, the case 9, from Figure 4.10 to Figure 4.12, has been generated in cold conditions and pulsating in shroud branch with $MFR = 0.7$. On the other hand, the case 43, from Figure 4.13 to Figure 4.15 has been generated pulsating in the hub branch with $MFR = 0.3$ and hot conditions. The equivalent engine speed of the excitation is 3900 rpm for the first case and 3700 rpm for the second case. Turbocharger speed is 97 krpm for both cases. The sound pressure level (SPL) from the pressure decomposition results for each turbine branch are presented in different graphs, confronting the experiments with the results provided by the model in both time and frequency domain. The label in the centre of each graph indicates the branch where the pulses are performed, the mass flow ratio and the conditions, following Table 5.1.

Regarding the nomenclature, incident and reflected are used for the pulsating branch reference the forward and backward travelling waves, i.e., towards and away from the turbine respectively. The same procedure has been used for the nomenclature of the other inlet branch, where indirect reflected and indirect transmitted have been used for the waves moving towards the turbine and away from the turbine respectively. In the outlet branch, the transmitted indicates the wave travelling downstream of the turbine in the direction of the flow, whereas the 2nd reflection wave travels from the outlet of the turbine in the opposite direction of the flow. This naming convention is also presented in Figure 4.4.

For performing the simulations, the incident component of the pressure is imposed in the pulsating inlet branch as a boundary condition along with the entropy level. In the same way, the indirect reflected wave is also imposed in the not pulsating inlet branch, whereas the 2nd reflection component is imposed in the outlet branch. In this way, for performing the validation, the capability of the model to predict the reflected and indirect reflected in the inlets, and the transmitted wave in the outlet is evaluated.

In each graph, the time domain result is aimed to check that the geometry of the turbine is correctly modelled and, in consequence, the shape and amplitude of the excitation is well predicted in terms of how it is evolving when it interacts with the different components of the turbine. Nevertheless, results in the time domain do not give easily identifiable information about how the pulse evolves in the turbine in the medium and high frequency spectra. Because of that, the equivalent frequency domain result has been obtained using Fourier's transformation applying a Hann [103] window to a Welch's average periodogram [102]. The frequency domain results are shown in terms of sound pressure level, computed with a reference pressure equal to 0.02 mPa. This way, what the bottom part of the graph is presenting is the amplitude in dB of each harmonic of the equivalent engine.

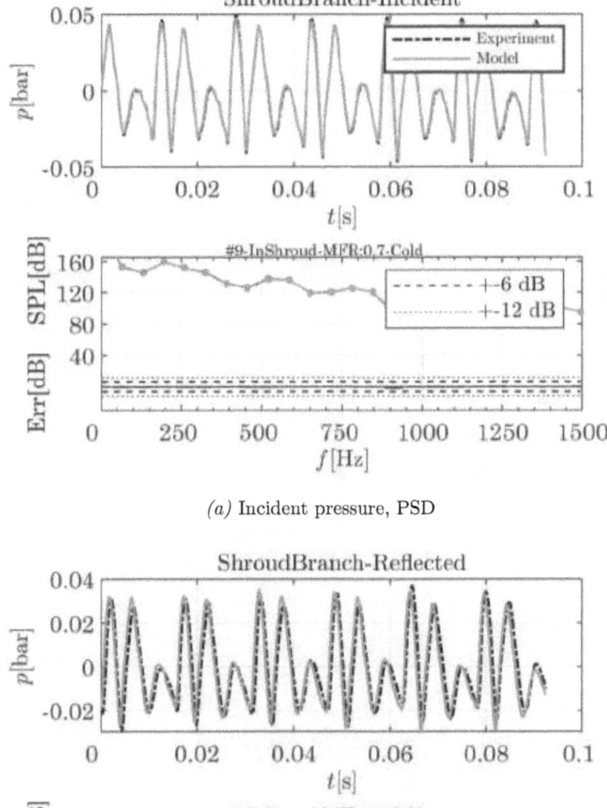

(a) Incident pressure, PSD

(b) Reflected pressure, PSD

Figure 4.10: Case 9 results: Incident is imposed (a), Reflected is computed (b).

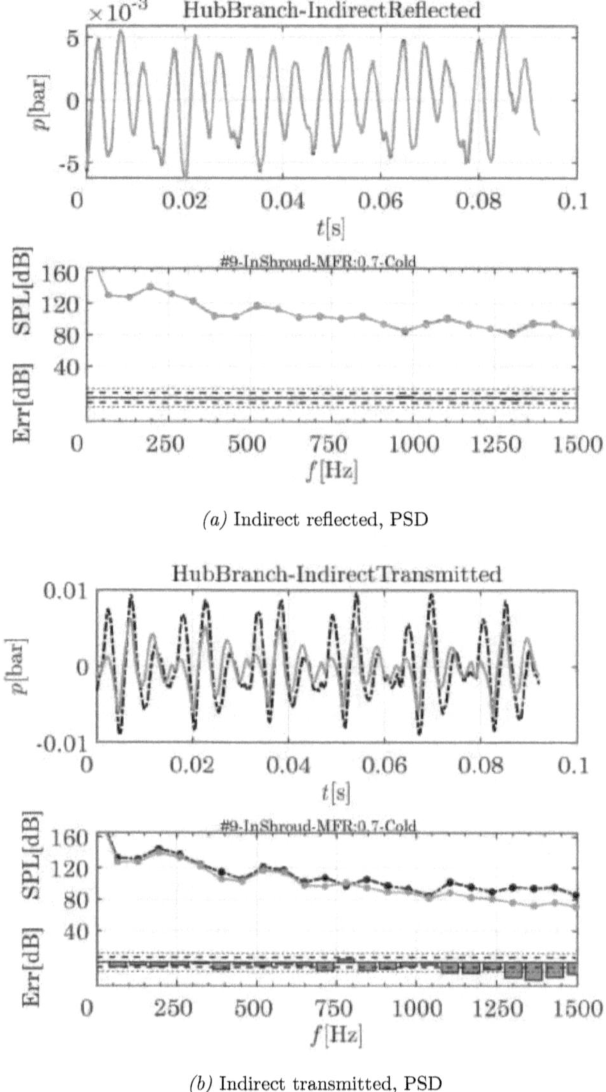

(a) Indirect reflected, PSD

(b) Indirect transmitted, PSD

Figure 4.11: Case 9 results: Ind. Reflected is imposed (a), Ind. Transmitted is computed (b).

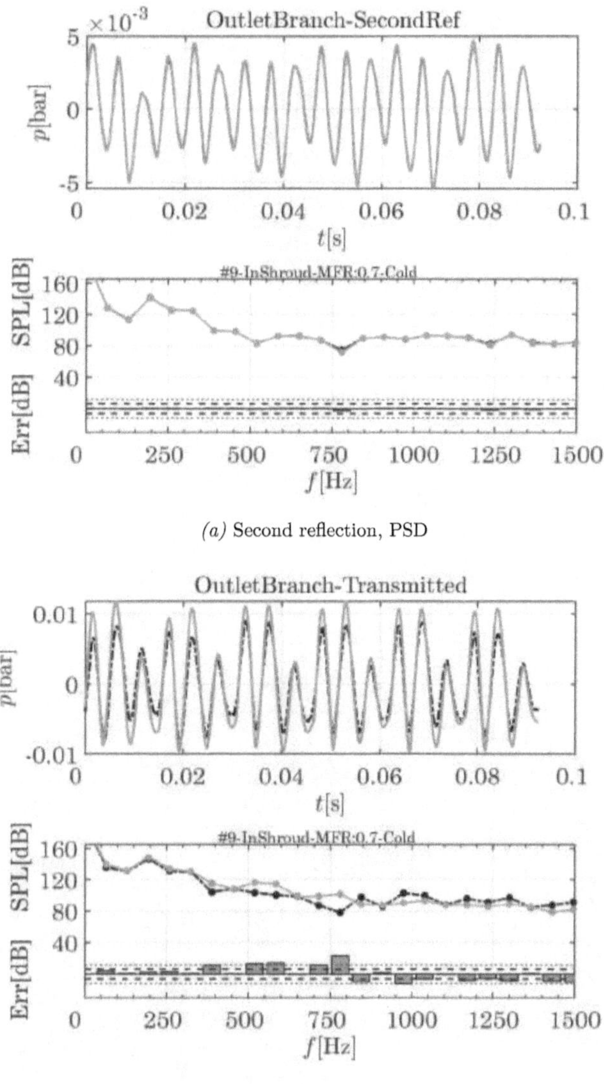

(a) Second reflection, PSD

(b) transmitted, PSD

Figure 4.12: Case 9 results: 2nd reflection is imposed (a), Transmitted is computed (b).

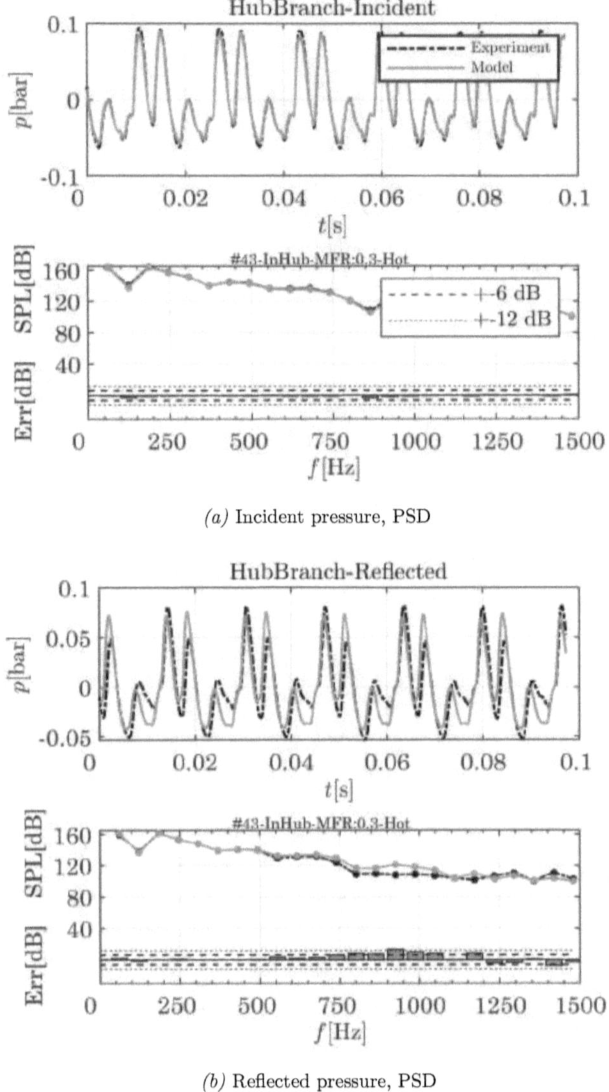

(a) Incident pressure, PSD

(b) Reflected pressure, PSD

Figure 4.13: Case 43 results: Incident is imposed (a), Reflected is computed (b).

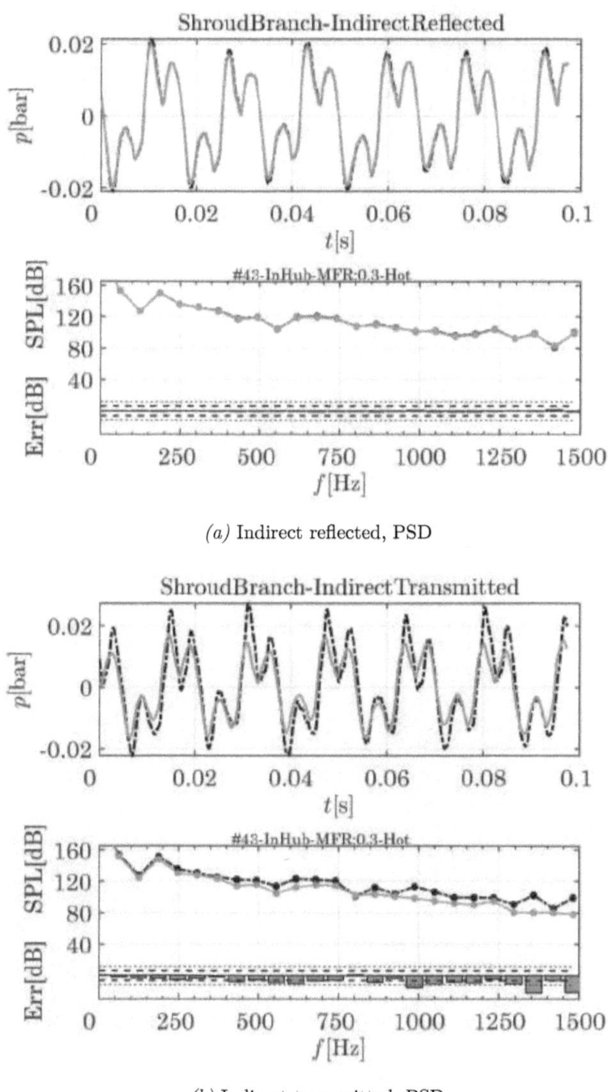

(a) Indirect reflected, PSD

(b) Indirect transmitted, PSD

Figure 4.14: Case 43 results: Ind. Reflected is imposed (a), Ind. Transmitted is computed (b).

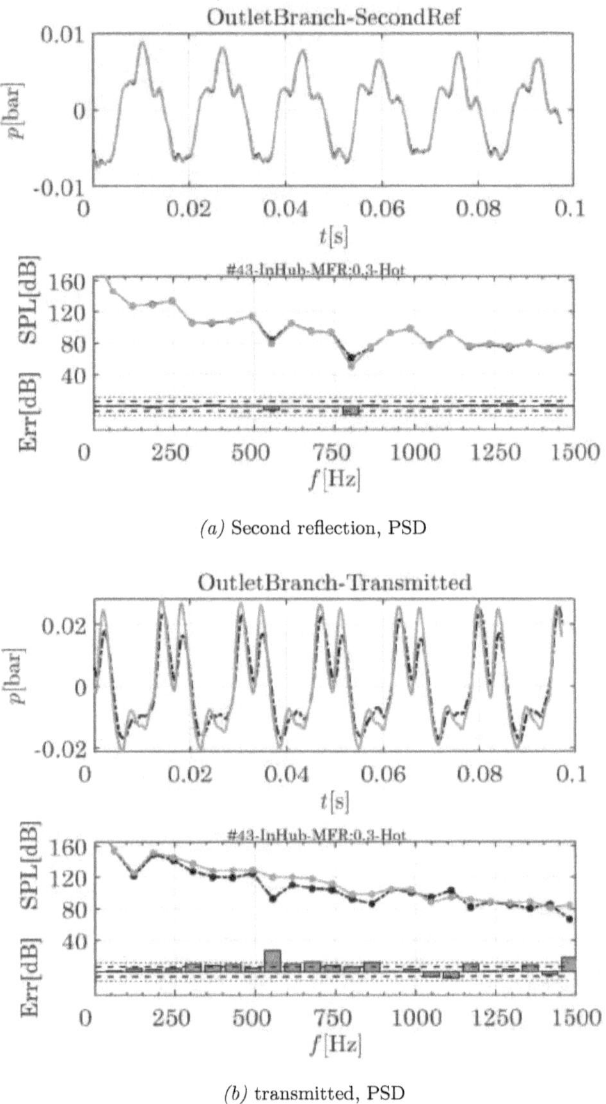

(a) Second reflection, PSD

(b) transmitted, PSD

Figure 4.15: Case 43 results: 2nd reflection is imposed (a), Transmitted is computed (b).

Figures: 4.10(a), 4.11(a), 4.12(a), 4.13(a), 4.14(a) and 4.15(a) show the pressure waves that are directly imposed as boundary conditions, whereas in 4.10(b), 4.11(b), 4.12(b), 4.13(b), 4.14(b) and 4.15(b) the pressure decomposition components are computed. Focusing in the pulsating branch results presented in 4.10(b) and 4.13(b), it can be observed how the reflected component is approximately conserving the shape of the incident, although the amplitude decreases, which is expected due to the attenuation produced once the excitation hits the turbine. The difference in amplitude between both cases is similar to what is found in single-entry turbines [1], being the amplitude of the reflected wave around half of that of the incident pressure wave. Continuing in 4.10(b) and 4.13(b), it is noteworthy the capability of the model to replicate the reflected component even in the high frequency domain, maintaining the error lower than 6 dB until almost 1000 Hz for both cases.

Regarding the other inlet branch, in which the rotating valve is not actuating, some relevant facts can be pointed out from the results obtained. It is interesting to analyse how the wave is transformed when it hits the rotor and travels backwards through the other inlet branch and how this is predicted by the model. To illustrate this with an example, in 4.10(b) pulses are performed in the shroud branch, so what the indirect transmitted graph represents in 4.11(b) is the effect that this excitation has produced in the hub branch. From the amplitude of the indirect transmission component it can be inferred that the amplitude has been considerably attenuated when compared to the original incident signal, going from the original amplitude of around 10 kPa in the incident wave of the hot case (4.13(a)), to approximately 2 kPa of the indirect transmitted component (4.14(b)). Nevertheless, the results indicate that, although attenuated, when operating in real engine conditions, the effect of the pulses that will travel to the not pulsating inlet branch are not negligible and might certainly affect the global performance of the engine. Model prediction for the indirect transmitted is not as extremely accurate as in the reflected component, but error is still maintained under the 6 dB level for the first half of the studied frequency range.

Finally, analysing the effect of the wave that travels through the turbine all the way to the outlet branch, the conclusions are similar than the ones exposed when analysing the indirect transmitted component. The amplitude is also considerably attenuated until the 2 kPa level (4.12(b) and 4.15(b)), but again is not negligible. This implies that when operating in normal conditions, non-linear pulses will cross the turbine at an still important fraction of their level at the inlet and find their way reaching the aftertreatment line. As it can be observed from the above mentioned figures, the model is still predicting reasonably accurately until medium and high frequencies. The errors in the prediction of the transmitted are however slightly higher, between the 6 dB and the 12 dB level. Some of the differences are expected due to the geometrical simplifications used in the model. Also, the experimental uncertainty is higher at the turbine outlet, as the amplitude of the pulses are smaller and it is more complicated to isolate the actual pressure signal from the electromagnetic and thermal noise of the transducers.

113

These two operating points are representative enough of the general performance of the model, nevertheless, a summary of the results obtained for the complete experimental campaign is showed in Figure 4.16. In this figure, the Root Mean Square Error (RMSE) of the first 8 harmonics is calculated for the 48 operating points, and plotted against the amplitude. From this result it can be inferred that the discrepancy between the model and the experiments is lower when the amplitude increases, what is consistent with a reduction of the experimental uncertainty with higher amplitude signals.

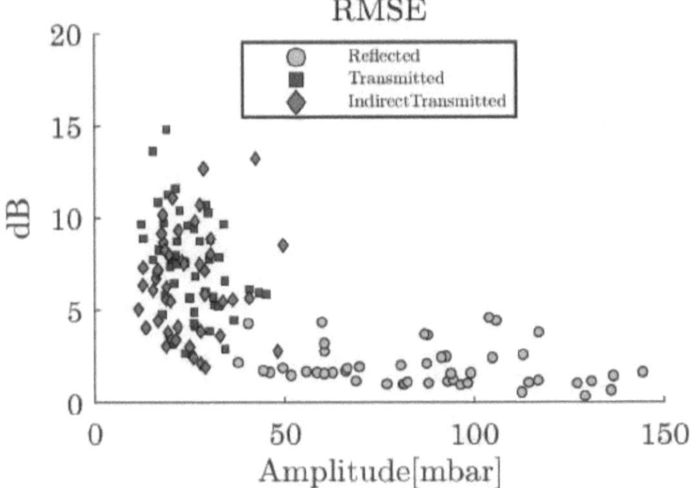

Figure 4.16: Root mean square error of the model for the harmonics in all the experimental points.

A pressure decomposition study is nevertheless, only one of several applications for an integrated turbine model with non-linear prediction capabilities. It is worth studying also some other variables, such as the turbine outlet temperature. In Figure 4.17 the turbine outlet temperature drop of each operating point of the experiment is presented, i.e., the difference between the maximum and the minimum instantaneous temperature during the pulse for the 48 operating points.

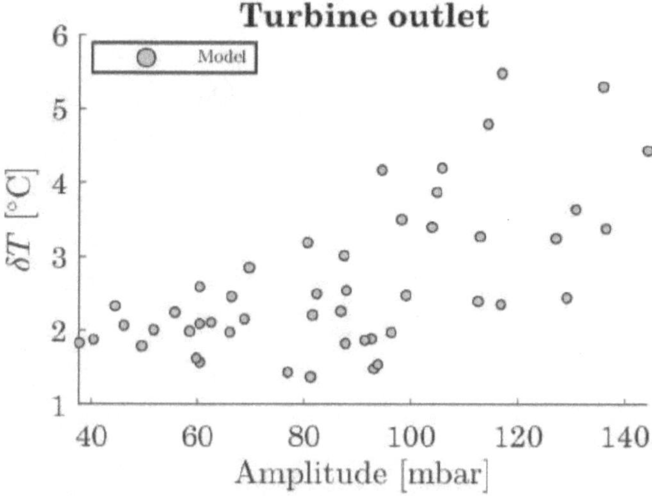

Figure 4.17: Temperature drop in turbine outlet during a pressure pulse.

From Figure 4.17 it can be stated that high amplitude pulses are generating high oscillations in the outlet temperature with almost 6 K for the higher amplitude points which, when dealing with higher real engine pulses will probably get over 10 K. Fluctuations in the outlet temperature might affect the exhaust line and the aftertreatment.

4.3.3 Apparent efficiency study

Due to the difficulties in measuring the individual outlet temperature for each turbine branch, it becomes necessary to define an efficiency based on the mixed outlet temperatures of both turbine entries. For such purposes, the apparent efficiency can be analytically expressed as in Equation 4.4 and Equation 4.5, which are obtained applying the diagram from Figure 4.18, explained in detail in previous works[129].

$$\eta_{\text{shroud}} = \frac{T_{3t,\text{shroud}} - T_{5t,\text{shroud}}}{T_{3t,\text{shroud}} - T_{5s,\text{shroud}}} \qquad (4.4)$$

$$\eta_{\text{hub}} = \frac{T_{3t,\text{hub}} - T_{5t,\text{hub}}}{T_{3t,\text{hub}} - T_{5s,\text{hub}}} \qquad (4.5)$$

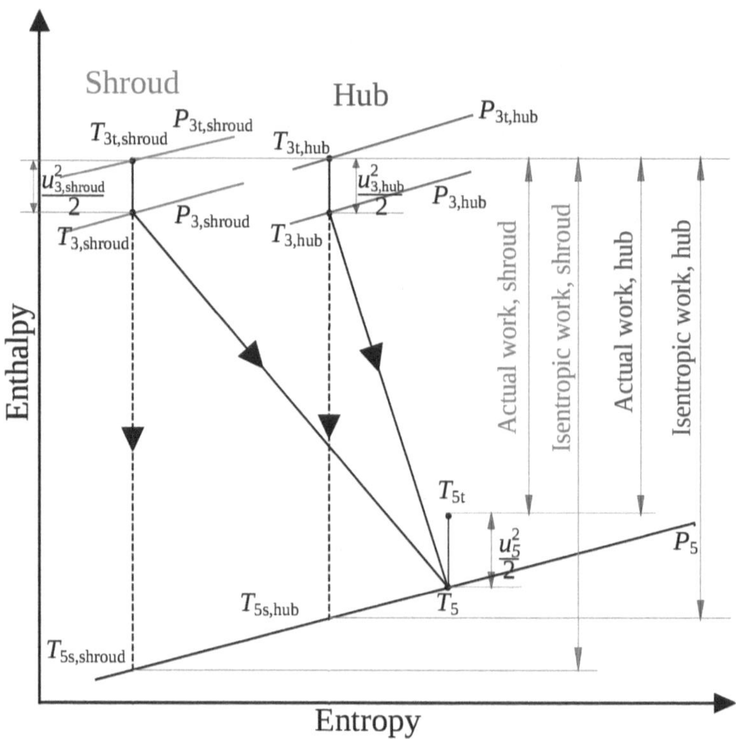

Figure 4.18: Enthalpy-entropy chart for the expansion process.

The objective of this model is not only to predict the non-linear acoustics, but to help taking better decisions during the design stage and calibration process of the turbocharger and its correspondent engine. In this sense, it is interesting to analyse the instantaneous variation of the efficiency, along with the mass flow ratio and the blade speed ratio. This can be used to give information about the interference effects between branches in terms of actual mass flow and power output of the turbine.

In Figure 4.19 and Figure 4.20, the instantaneous result of the apparent efficiency of the previous example cases after 1 s of simulation is presented as the independent variable, using the mass flow ratio and the blade speed ratio as dependent variables.

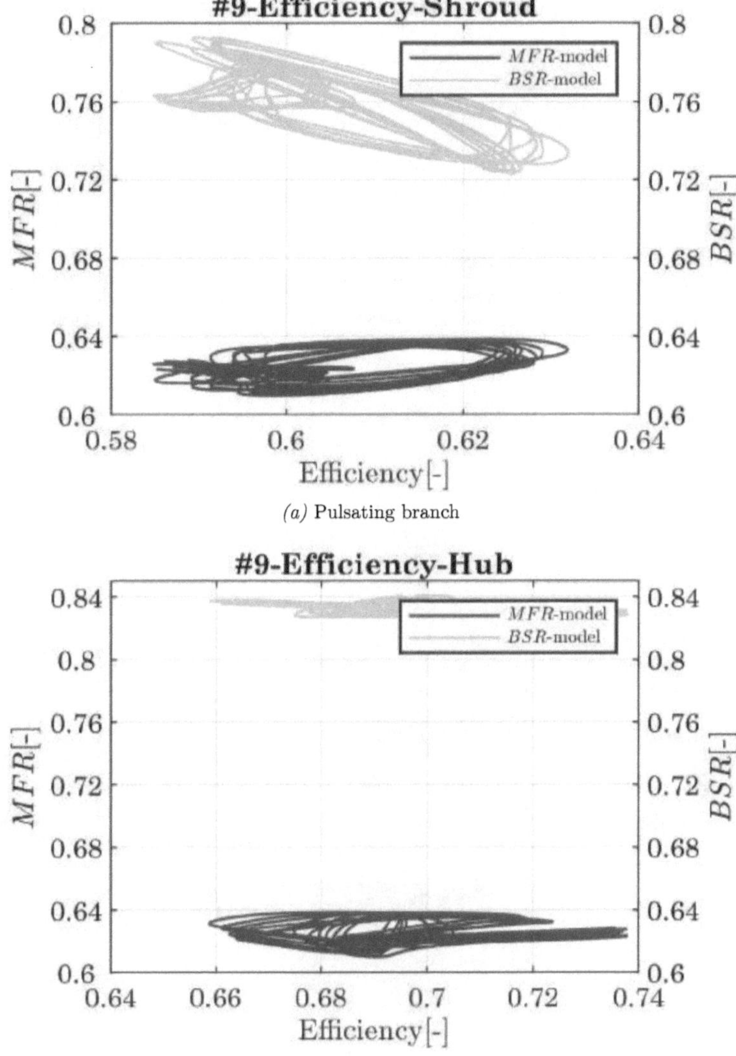

(a) Pulsating branch

(b) Not pulsating branch

Figure 4.19: Instantaneous apparent efficiency simulation results. Case 9 (a, b).

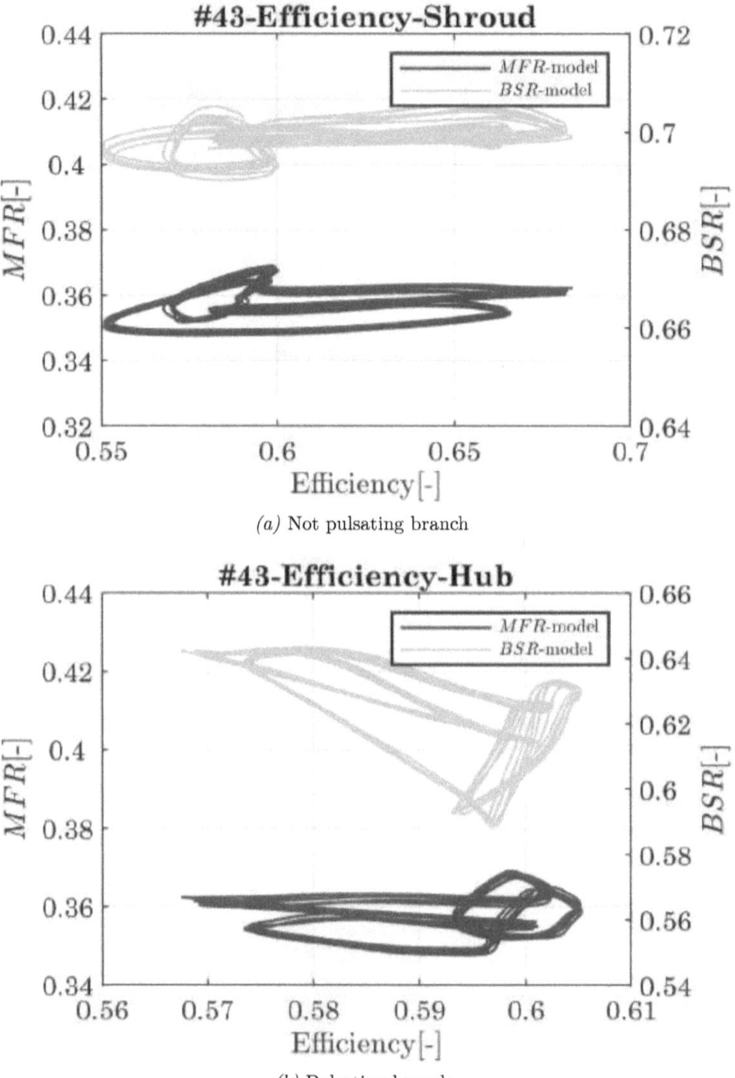

(a) Not pulsating branch

(b) Pulsating branch

Figure 4.20: Instantaneous apparent efficiency simulation results. Case 43 (a, b).

From this efficiency study, different points can be noted. Firstly, the range of the efficiency is wider in the not pulsating branch results (Figure 4.19(b) and Figure 4.20(a)), reaching a range of more than 0.12 points in shroud for case 43. In contrast, the blade speed ratio present its highest variation in the pulsating branch (Figure 4.19(a) and Figure 4.20(b)), from approximately 0.72 until almost 0.8. Finally, from the graphs can be inferred that the mass flow ratio presents a modest but not negligible variation, varying from 0.6 to 0.64 for case 9 and from 0.35 to 0.37 for case 43.

Interested findings can be inferred from the apparent efficiency simulations results above commented:

- Both turbine branches can not be studied as isolated entities to get their mass flow and power output, as there are interference effects between them.

- The apparent efficiency changes in the non-pulsating branch can be large due to the pulses produced in the other branch. In an engine, the branch connected to an exhaust valve that is closed will present these large efficiency changes while its mass flow drops.

- The mass flow ratio can present extra oscillations of moderate amplitude due to the interference of the pressure pulses between turbine inlet branches.

4.3.4   Instantaneous mass flow analysis

A model able to reproduce correctly the non-linear effects of the engine pulses in the turbine can also give valuable information in terms of mass flow. In particular, one of the most interesting analysis that can be carried out is to evaluate the effect that a pulse of the active branch produces in the not pulsating turbine branch. Not in vain, these oscillations travelling upstream of the turbine through the engine might very likely affect the volumetric efficiency.

In Figure 4.21, the mass flow amplitude is presented. This variable is obtained using the absolute difference between the maximum and the minimum value of the instantaneous mass flow during a pulse for each operating point. In the x axis, the mass flow amplitude reached in the branch where the pulses are performed is presented. On the other hand, the dependent variable is obtained with the quotient between the mass flow amplitude in the not pulsating branch and the mass flow amplitude in the pulsating branch. Thus, this quotient indicates which percentage of the pulse in terms of mass flow is reaching the not pulsating or "passive" branch due to the pulses generated in the active branch.

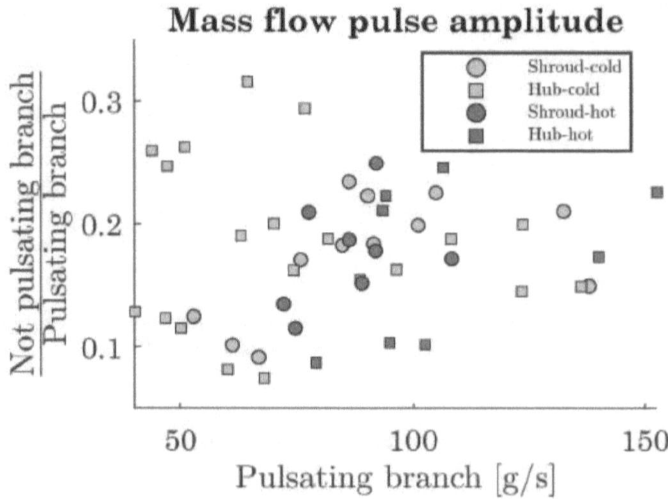

Figure 4.21: Amplitude of the instantaneous mass flow.

In view of these results, some considerations can be done:

- The effects that the pulses produce in the passive branch in terms of mass flow are probably sufficiently large to affect the performance, specially when the turbine faces real engine pulses generated during transients. In fact, for most of the points, between the 20 % and 30 % of the oscillation manage to travel from the turbine pulsating branch until the outlet and all the way back through the other inlet branch.

- The hot conditions produce bigger amplitude pulses, reaching the level of around $150\,\mathrm{g\,s^{-1}}$ or more. These pulses will lead to bigger transmitted components to the not pulsating branch. The relative effect in the passive branch is not, however, significantly different from the effect produced in the cold cases, as it can be inferred from the values of the quotient between amplitudes.

- For some points in the lower amplitude range (lower than $100\,\mathrm{g\,s^{-1}}$), the pulses produced in the hub entry are better transmitted to the other branch that the ones produced in the shroud branch. This could be one of the effects generated by the asymmetry of the turbine entries. Conclusions can vary when analysing other turbochargers, however, the main objective of this section was to illustrate the potential of the model.

After presenting a global view of the mass flow, it is worth analysing the capability of the model to reproduce the behaviour of the instantaneous mass flow in the time domain. Following the philosophy of chapter 3 for the selection of the cases, in Figure 4.22 the instantaneous mass flow evolution is shown for similar pulsating conditions but pulsating in shroud branch (4.22(a) and 4.22(c)) and pulsating in hub branch (4.22(b) and 4.22(d)). From the results it can be observed how the imposed pulses are very similar in amplitude and shape regardless of the branch in which the pulses are performed, reaching values in the approximate range of 20 g/s and 120 g/s for the pulsating branch and between 60 g/s and 80 g/s for the passive branch (reaching 85 g/s for case #29).

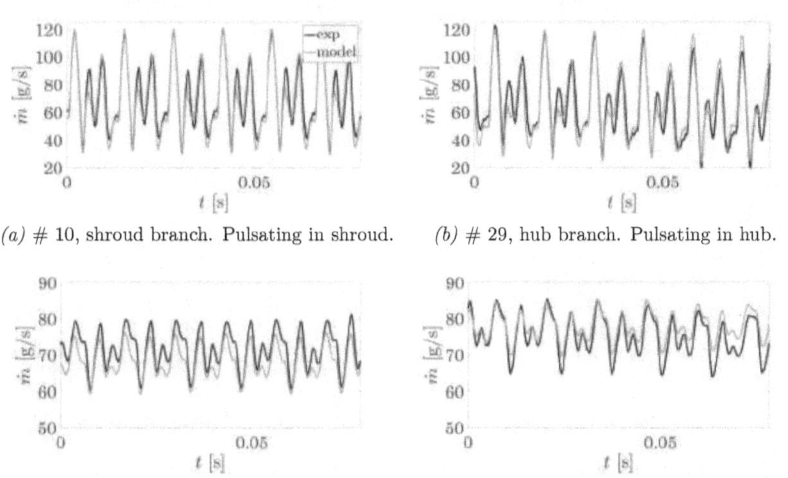

(a) # 10, shroud branch. Pulsating in shroud.   (b) # 29, hub branch. Pulsating in hub.

(c) #10, hub branch. Pulsating in shroud.   (d) #29, shroud branch. Pulsating in hub.

Figure 4.22: Instantaneous mass flow for pulsating branch for case #10 (a) and case #29 (b) and passive branch for case #10 (c) and case #29 (d). Both cases are the same conditions: MFR 0.5, 4600 rpm and cold flow.

Furthermore, the model is able to capture the instantaneous mass flow evolution fairly accurately in the majority of the cases, which is coherent with the ability to reproduce the first pressure harmonic, as it can be observed from Figure 4.10 to Figure 4.15. However, there were some exceptions in cases in which the model presented higher discrepancies with the experiments. In particular, when focusing in some cases with low engine equivalent speed, the model seemed to correctly capture the shape of the mass flow traces, but with a reduced amplitude, as it can be observed in the hub result (4.23(b)) of Figure 4.23. The fact that this rare behaviour appears only for the lower engine equivalent speed, suggests the idea that in those conditions a resonance might occur due to the simplification of the outlet geometry. Not in vain, the outlet of the rotor is modelled as a simple straight pipe, but the real geometry includes asymmetries also due to the waste-gate, that might induce to gradient velocities near the walls, which might be particularly relevant at lower velocities. Despite this tricky phenomenon, the model is able to correctly reproduce the shape and amplitude of the instantaneous mass flow for the majority of the cases.

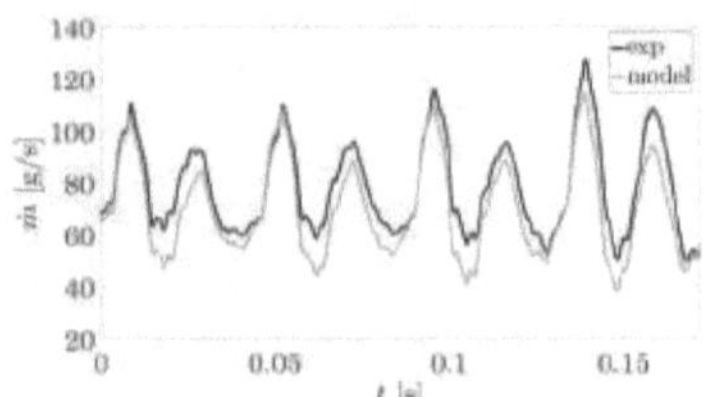

(a) # 1, shroud branch. Pulsating in shroud.

(b) # 17, hub branch. Pulsating in hub.

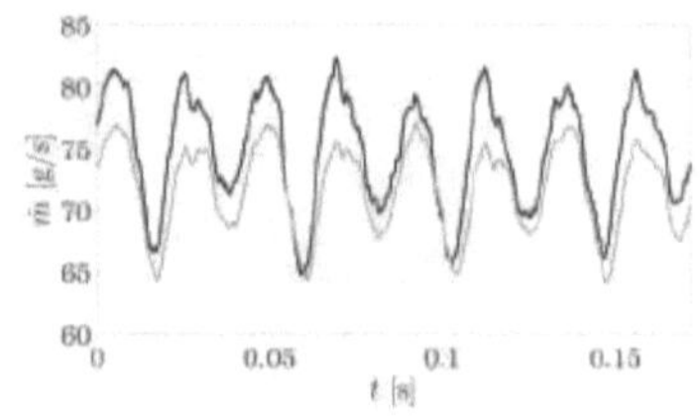

(c) #1, hub branch. Pulsating in shroud.

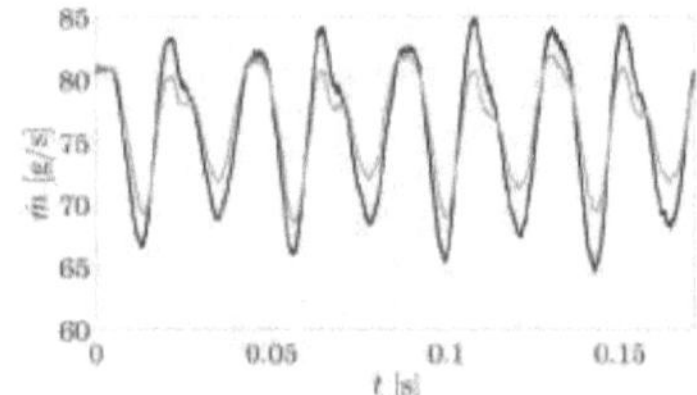

(d) #17, shroud branch. Pulsating in hub.

Figure 4.23: Instantaneous mass flow for pulsating branch for case #1 (a) and case #17 (b) and passive branch for case #1 (c) and case #17 (d). Both cases are the same conditions: MFR 0.5, 1400 rpm and cold flow.

4.4   Conclusions

In this chapter, a 1D modelling method to predict the two-scroll turbine performance under non-linear pulsating conditions is presented. The model has been validated against experimental data for several simulated engine conditions, i.e., full and partial admission conditions, different equivalent engine speeds, and pulsating in both inlet branches separately for cold and hot flow conditions.

From the pressure decomposition study, it can be inferred that the model is able to capture the non-linear effects until the medium and sometimes high frequency spectra. The model is less effective when the amplitude is lower, probably due to the higher effect that the experimental uncertainty is producing for these cases. A good level of accuracy is expected when trying to predict real engine conditions.

Although instantaneous experimental results were not obtained for the efficiency due to the difficulties that it implies, the study based on simulation presented indicates that instantaneous variation in the efficiency is significant enough in all the turbine branches to have an impact in the turbine performance. The mass flow ratio range of operation in the contrary is narrower, which opens the possibility of studying different averaged or filtered versions of this parameter without affecting the prediction performance of the model.

From these results it is clear that during the engine exhaust process, interferences are still transmitted between the different turbine branches that will affect the performance of the turbine. The correct modelling of the increasingly used two-scroll turbine under realistic conditions could be very useful when analysing the turbocharger matching, as the low frequency oscillations travelling upstream of the turbine will affect the engine volumetric efficiency and even the turbocharger operating point. Furthermore, a better prediction of the medium and high frequency pulses that are travelling through the exhaust line could lead to a more efficient design of the exhaust line and the muffler. A good prediction of the turbine outlet temperature oscillations due to the pressure pulses should also enable to better compute the aftertreatment behaviour.

Chapter 4 References

[1]   J. Galindo, F. J. Arnau, L. M. García-Cuevas, and P. Soler. "Experimental validation of a quasi-two-dimensional radial turbine model". *International Journal of Engine Research* (2018). ISSN: 1468-0874. DOI: 10.1177/1468087418788502 (cit. on pp. xi, 8, 57, 62, 84, 90, 113, 131, 133, 139, 177).

[2]   J. Serrano, F. Arnau, L. M. García-Cuevas, P. Soler, L. Smith, R. Cheung, and B. Pla. "An Experimental Method to Test Twin and Double Entry Automotive Turbines in Realistic Engine Pulse Conditions". In: *WCX SAE World Congress Experience*. SAE Technical Paper 2019-01-0319. 2019. DOI: 10.4271/2019-01-0319 (cit. on pp. xi, 11, 100, 130).

[17]   CMT – Motores T'ermicos, Universitat Politècnica de València. *OpenWAM*. 2016. URL: http://www.openwam.org/ (cit. on pp. 7, 11, 22, 98, 133, 178).

[23]   F. Payri, J. Benajes, and M. Reyes. "Modelling of supercharger turbines in internal-combustion engines". *International Journal of Mechanical Sciences* 38(8) (1996), pp. 853 –869. ISSN: 0020-7403. DOI: 10.1016/0020-7403(95)00105-0 (cit. on pp. 8, 91).

[24]   H. Chen and D. Winterbone. "A method to predict performance of vaneless radial turbine under steady and unsteady flow conditions". In: *Turbocharging*

and Turbochargers. Institution of Mechanical Engineers. 1990, pp. 13–22 (cit. on pp. 8, 20, 21, 97).

[25] J. R. Serrano, F. J. Arnau, V. Dolz, A. Tiseira, and C. Cervelló. "A model of turbocharger radial turbines appropriate to be used in zero- and one-dimensional gas dynamics codes for internal combustion engines modelling". *Energy Conversion and Management* 49(12) (2008), pp. 3729 –3745. ISSN: 0196-8904. DOI: 10.1016/j.enconman.2008.06.031 (cit. on pp. 8, 30, 31, 96, 99, 100, 134).

[27] F. Payri, P. Olmeda, F. J. Arnau, A. Dombrovsky, and L. Smith. "External heat losses in small turbochargers: Model and experiments". *Energy* 71 (2014), pp. 534 –546. ISSN: 0360-5442. DOI: 10.1016/j.energy.2014.04.096 (cit. on pp. 8, 20, 27, 92, 131).

[28] J. R. Serrano, P. Olmeda, F. J. Arnau, A. Dombrovsky, and L. Smith. "Turbocharger heat transfer and mechanical losses influence in predicting engines performance by using one-dimensional simulation codes". *Energy* 86 (2015), pp. 204 –218. DOI: 10.1016/j.energy.2015.03.130 (cit. on pp. 8, 27, 92, 131).

[29] A. Gil, A. Tiseira, L. M. García-Cuevas, T. Rodríguez Usaquén, and G. Mijotte. "Fast three-dimensional heat transfer model for computing internal temperatures in the bearing housing of automotive turbochargers". *International Journal of Engine Research* (2018). DOI: 10.1177/1468087418804949 (cit. on pp. 8, 27, 92, 131).

[30] J. R. Serrano, F. J. Arnau, L. M. García-Cuevas, A. Dombrovsky, and H. Tartoussi. "Development and validation of a radial turbine efficiency and mass flow model at design and off-design conditions". *Energy Conversion and Management* 128 (2016), pp. 281 –293. ISSN: 0196-8904. DOI: 10.1016/j.enconman.2016.09.032 (cit. on pp. 8, 26–29, 96, 98, 131, 134, 135, 139, 157, 178).

[31] J. Galindo, R. Navarro, L. M. García-Cuevas, D. Tarí, H. Tartoussi, and S. Guilain. "A zonal approach for estimating pressure ratio at compressor extreme off-design conditions". *International Journal of Engine Research* 20(4) (2019), pp. 393–404. DOI: 10.1177/1468087418754899 (cit. on pp. 8, 92, 131).

[32] J. R. Serrano, P. Olmeda, A. Tiseira, L. M. García-Cuevas, and A. Lefebvre. "Theoretical and experimental study of mechanical losses in automotive turbochargers". *Energy* 55(0) (2013), pp. 888 –898. ISSN: 0360-5442. DOI: 10.1016/j.energy.2013.04.042 (cit. on pp. 8, 21, 92, 131).

[34] J. Serrano, F. Arnau, L. Garcéa-Cuevas, and V. Samala. "Development of flow oriented model for extrapolation and interpolation off-design performance of twin-entry and dual volute radial-inflow turbines working under different flow admission conditions". *Energy* (). Under review (cit. on pp. 8, 92, 99, 100, 177).

[36] A. W. Costall, R. M. McDavid, R. F. Martinez-Botas, and N. C. Baines. "Pulse Performance Modeling of a Twin Entry Turbocharger Turbine Under Full and Unequal Admission". *Journal of Turbomachinery* 133(2) (2011), p. 021005. ISSN: 0889504X. DOI: 10.1115/1.4000566 (cit. on pp. 8, 91).

[38]   M. S. Chiong, S. Rajoo, R. F. Martinez-Botas, and A. W. Costall. "Engine turbocharger performance prediction: One-dimensional modeling of a twin entry turbine". *Energy Conversion and Management* 57 (2012), pp. 68–78. ISSN: 01968904. DOI: 10.1016/j.enconman.2011.12.001 (cit. on pp. 8, 91).

[39]   M. S. Chiong, S. Rajoo, A. Romagnoli, A. W. Costall, and R. F. Martinez-Botas. "Assessment of Partial-Admission Characteristics in Twin-Entry Turbine Pulse Performance Modelling". In: *Volume 2C: Turbomachinery*. 2015, V02CT42A022. ISBN: 978-0-7918-5665-9. DOI: 10.1115/GT2015-42687 (cit. on pp. 8, 24, 92).

[42]   S. Rajoo, A. Romagnoli, and R. F. Martinez-Botas. "Unsteady performance analysis of a twin-entry variable geometry turbocharger turbine". *Energy* 38(1) (2012), pp. 176–189. ISSN: 03605442. DOI: 10.1016/j.energy.2011.12.017 (cit. on pp. 8, 57, 91).

[43]   S. Rajoo and R. F. Martinez-Botas. "Variable Geometry Mixed Flow Turbine for Turbochargers: An Experimental Study". *International Journal of Fluid Machinery and Systems* 1(1) (2008), pp. 155–168. ISSN: 1882-9554. DOI: 10.5293/IJFMS.2008.1.1.155 (cit. on pp. 8, 91).

[45]   C. D. Copeland, R. F. Martinez-Botas, and M. Seiler. "Unsteady Performance of a Double Entry Turbocharger Turbine With a Comparison to Steady Flow Conditions". *Journal of Turbomachinery* 134(2) (2012), p. 021022. ISSN: 0889504X. DOI: 10.1115/1.4003171 (cit. on pp. 8, 91).

[67]   J. R. Serrano, P. Olmeda, A. Páez, and F. Vidal. "An experimental procedure to determine heat transfer properties of turbochargers". *Measurement Science and Technology* 21(3) (2010), p. 035109. DOI: 10.1088/0957-0233/21/3/035109 (cit. on pp. 20, 34, 95).

[79]   J. Galindo, P. Fajardo, R. Navarro, and L. M. García-Cuevas. "Characterization of a radial turbocharger turbine in pulsating flow by means of CFD and its application to engine modeling". *Applied Energy* 103(0) (2013), pp. 116 –127. ISSN: 0306-2619. DOI: 10.1016/j.apenergy.2012.09.013 (cit. on pp. 21, 56, 90, 130).

[86]   K. G. Hohenberg, P. J. Newton, R. F. Martinez-Botas, M. Halamek, K. Maeda, and J. Bouilly. "Development and Experimental Validation of a Low Order Turbine Model Under Highly Pulsating Flow". In: *Turbo Expo: Power for Land, Sea, and Air, Volume 2C: Turbomachinery*. ASME. June 2017, V02CT44A023. DOI: 10.1115/GT2017-63983 (cit. on pp. 21, 57, 90, 130).

[87]   J. R. Serrano, A. Tiseira, L. M. García-Cuevas, L. Inhestern, and H. Tartoussi. "Radial turbine performance measurement under extreme off-design conditions". *Energy* 125 (2017), pp. 72–84. DOI: 10.1016/j.energy.2017.02.118 (cit. on pp. 21, 57, 90, 135).

[88]   A. J. Torregrosa, A. Broatch, R. Navarro, and J. García-Tíscar. "Acoustic characterization of automotive turbocompressors". *International Journal of Engine Research* 16(1) (2015), pp. 31–37. DOI: 10.1177/1468087414562866 (cit. on pp. 21, 57, 90, 130).

[102] P. Welch. "The Use of Fast Fourier Transform for the Estimation of Power Spectra: A Method Based on Time Averaging Over Short, Modified Periodograms". *IEEE Transactions on Audio Electroacoustics* 15 (2 1967), pp. 70–73. ISSN: 0018-9278. DOI: 10.1109/TAU.1967.1161901 (cit. on pp. 39, 106, 136).

[103] F. Harris. "On the use of windows for harmonic analysis with the discrete Fourier transform". In: *Proceedings of the IEEE*. Vol. 66. IEEE, 1978, pp. 51–83. DOI: 10.1109/PROC.1978.10837 (cit. on pp. 39, 106).

[104] R. Hogg. "Life beyond euro VI". *automotiveworld* (2014). URL: http://www.automotiveworld.com/megatrends-articles/life-beyond-euro-vi/ (cit. on pp. 56, 90).

[105] M. Knopf. "How low can we go? Downsizing the internal combustion engine". *Ingenia* (2011). URL: https://www.ingenia.org.uk/Ingenia/Articles/11612c67-32db-477a-af8f-38cef44ee7b2 (cit. on pp. 56, 90).

[108] A. Torregrosa, J. Serrano, J. Dopazo, and S. Soltani. "Experiments on wave transmission and reflection by turbochargers in engine operating conditions". *SAE Technical Papers* 2006(01:0022) (2006). DOI: 10.4271/2006-01-0022 (cit. on pp. 57, 90).

[114] M. Yang, R. F. Martinez-Botas, S. Rajoo, T. Yokoyama, and S. Ibaraki. "Influence of Volute Cross-Sectional Shape of a Nozzleless Turbocharger Turbine Under Pulsating Flow Conditions". In: *Volume 2D: Turbomachinery*. ASME, 2014, V02DT42A025. ISBN: 978-0-7918-4563-9. DOI: 10.1115/GT2014-26150 (cit. on pp. 57, 91).

[115] M. Cerdoun and A. Ghenaiet. "Unsteady behaviour of a twin entry radial turbine under engine like inlet flow conditions". *Applied Thermal Engineering* 130 (2018), pp. 93–111. ISSN: 1359-4311. DOI: 10.1016/j.applthermaleng.2017.11.001 (cit. on pp. 57, 91).

[117] R. Kabral, Y. El Nemr, C. Ludwig, R. Mirlach, P. Koutsovasilis, A. Masrane, and M. Åbom. "Experimental acoustic characterization of automotive twin-scroll turbine". In: *12th European Conference on Turbomachinery Fluid Dynamics and Thermodynamics ETC*. 2017, p. 363 (cit. on pp. 57, 75, 92).

[118] N. Watson and M. Janota. "Turbocharging the internal combustion engine". *The Macmillan Press Ltd* (1982) (cit. on p. 90).

[119] D. Winterbone, B Nikpour, and H Frost. "A contribution to the understanding of turbocharger turbine performance in pulsating flow". *Inst Mech Eng Semin, paper C433011* (1991) (cit. on p. 90).

[120] D. Winterbone and R. Pearson. "Turbocharger turbine performance under unsteady flow-a review of experimental results and proposed models". *Inst Mech Eng Semin, paper C55403198* (1998) (cit. on pp. 90, 130).

[121] F. Piscaglia, A. Onorati, S. Marelli, and M. Capobianco. "A detailed one-dimensional model to predict the unsteady behavior of turbocharger turbines for internal combustion engine applications". *International Journal of Engine Research* 20(3) (2019), pp. 327–349. DOI: 10.1177/1468087417752525 (cit. on pp. 90, 130).

[122] N. Cappelaere, A. Dazin, and G. Bois. "An industrial experimental methodology for the unsteady characterization of automotive turbocharger". *16th International Symposium on Transport Phe- nomena and Dynamics of Rotating Machinery* (2016) (cit. on p. 91).

[123] N. Baines, A. Hajilouy-Benisi, and J. Yeo. "The pulse flow performance and modelling of radial inflow turbines". In: *5th International Conference on Turbochargers, Inst Mech Eng, London, Paper C48400694.* 1994, pp. 209–219 (cit. on p. 91).

[124] C. D. Copeland, R. F. Martinez-Botas, and M. Seiler. "Comparison Between Steady and Unsteady Double-Entry Turbine Performance Using the Quasi-Steady Assumption". *Journal of Turbomachinery* 133(3) (2011), p. 031001. ISSN: 0889504X. DOI: 10.1115/1.4000580 (cit. on p. 91).

[125] M. Yang, R. F. Martinez-Botas, S. Rajoo, T. Yokoyama, and S. Ibaraki. "An investigation of volute cross-sectional shape on turbocharger turbine under pulsating conditions in internal combustion engine". *Energy Conversion and Management* 105 (2015). ISSN: 01968904. DOI: 10.1016/j.enconman.2015.06.038 (cit. on p. 91).

[126] C. D. Copeland, P. J. Newton, R. F. Martinez-Botas, and M. Seiler. "The Effect of Unequal Admission on the Performance and Loss Generation in a Double-Entry Turbocharger Turbine". *Journal of Turbomachinery* 134(2) (2012), p. 021004. ISSN: 0889504X. DOI: 10.1115/1.4003226 (cit. on p. 91).

[127] J. Macek, Z. Zak, and O. Vitek. "Physical Model of a Twin-scroll Turbine with Unsteady Flow". *SAE Technical Papers* (2015). DOI: 10.4271/2015-01-1718 (cit. on p. 91).

[128] M. S. Chiong, S. Rajoo, A. Romagnoli, A. W. Costall, and R. F. Martinez-Botas. "One-dimensional pulse-flow modeling of a twin-scroll turbine". *Energy* 115 (2016), pp. 1291–1304. ISSN: 03605442. DOI: 10.1016/j.energy.2016.09.041 (cit. on p. 92).

[129] J. Serrano, F. Arnau, L. García-Cuevas, V. Samala, and L. Smith. "Experimental approach for the characterization and performance analysis of twin entry radial-inflow turbines in a gas stand and with different flow admission conditions". *Applied Thermal Engineering* 159 (2019), p. 113737. DOI: 10.1016/j.applthermaleng.2019.113737 (cit. on pp. 92, 95, 98, 102, 115).

[130] R. Baar, C. Biet, V. Boxberger, H. Mai, and R. Zimmermann. "New Evaluation of Turbocharger Components based on Turbine Outlet Temperature Measurements in Adiabatic Conditions". In: *15th International Symposium on Transport Phenomena and Dynamics of Rotating Machinery*. 2014. 2014 (cit. on p. 95).

[131] R Zimmermann, R Baar, and C Biet. "Determination of the isentropic turbine efficiency due to adiabatic measurements and the validation of the conditions via a new criterion". *Proceedings of the Institution of Mechanical Engineers, art C: Journal of Mechanical Engineering Science* 232(24) (2018), pp. 4485–4494. DOI: 10.1177/0954406216670683 (cit. on p. 95).

[132] J. R. Serrano, F. J. Arnau, P. Fajardo, and M. A. Reyes-Belmonte. "Contribution to the Modeling and Understanding of Cold Pulsating Flow Influence in the Efficiency of Small Radial Turbines for Turbochargers". *Journal of Engineering for Gas Turbines and Power* 134(10) (2012) (cit. on pp. 96, 98).

[133] H. Chen, I. Hakeem, and R. F. Martínez-Botas. "Modelling of a turbocharger turbine under pulsating inlet conditions". In: *Proceedings of the Institution of Mechanical Engineers, Part A: Journal of Power and Energy*. Institution of Mechanical Engineers. Oct. 1996, pp. 397–408 (cit. on p. 97).

Chapter 5

A fast 1D model for turbine sound and noise characterisation

Contents

5.1 Introduction

The acoustic performance of radial turbochargers can be divided into passive and active categories [134]. In the compressor, active noise, including whoosh noise, is generated at a broad range of frequencies[90, 135]. Also, the geometry of the elements directly upstream or at the inlet of the compressor affects its active noise characteristics, as shown by Broatch et al. [136] or Sharma et al. [137] In the turbine, sound and noise are actively generated at high frequencies and are associated with the rotation of the rotor blades, including rotating shock waves at high turbine expansion ratios. Passive acoustic properties influence the low frequency engine pulsations and are affected by the geometry of the turbine and its operating conditions. Experimental measurements to obtain these passive properties can be conducted as described in Peat et al.[138], i.e., producing pressure pulses at the turbine inlet by means of an electrovalve and measuring the oscillations in predefined points in the turbine inlet and outlet ducts, while maintaining an almost-constant rotational speed and mass flow. With some adaptations, higher amplitude pulses may be used [139], obtaining the backward- and forward-travelling pressure waves with beamforming arrays at both the turbine inlet and outlet ducts. This method has been also used and validated with radial compressors [88], demonstrating that a three-sensor-phased array beamforming technique is a very reliable method in the plane wave frequency range. Also twin-scroll turbines have been characterised experimentally under linear acoustic excitations[140], or even non-linear acoustic excitations[2]. Those experimental characterisations may be prohibitive in a very broad range of operating conditions. However, the acoustic performance measurement provides very valuable information in order to develop and validate fast numerical models that may be used during engine development phases. High fidelity computational fluid dynamics (CFD) characterisations can provide valuable information, too. In the last five years, several studies about CFD characterisation of active noise generation in radial compressors can be found[135, 141, 142]. The literature about CFD acoustic and noise characterisation of radial turbines has been scarce, nevertheless, some notable exceptions can be found such as the work by Marsan and Moreu[143].

The models for predicting the performance of radial turbines, including their behaviour in pulsating flow conditions, have evolved during the last decades as the computational resources have improved. From the early models proposed by Wallace et al [144]or Winterbone and Pearson[120] to the newest additions proposed by Hohenberg et al. [86], Zhanming et al. [93] or Piscaglia et al. [121], they tend to focus on performance parameters such as the instantaneous flow rate or torque. Even high-fidelity three-dimensional simulations are routinely carried out in the present day, as in the work by Galindo et al.[79], where a radial turbine is simulated under pulsating flow conditions. Reasonably fast simulations are usually carried out with one-dimensional discretisations of the geometry of the turbine, using equivalent tapered ducts for the turbine inlet, volute and outlet, as described by Costall et al. [35]. Some improvements can be achieved against this simplification without a big impact in computational costs by improving the computation of the volute. The volute can be discretised as a one-dimensional tapered duct with a lateral connection to the vanes of the stator or the vaneless space before the rotor, depending on the turbine stator characteristics [91, 93, 145]. It is also worth mentioning the difficulties in correlating turbine power with experiments as found by Avola et al. [146]

To the knowledge of the author, the work focused in acoustic predictions has been limited, although there are some notable exceptions. Veloso et al. [147] propose an interesting approach for the modelling problem of the passive acoustic properties of radial turbines and compressors. In their work, they model the turbomachine as several one-dimensional tapered pipes connected together, with the rotor being a constricting pipe. The pressure and temperature drop in the turbine was set by means of PID controllers to match experimental data, and the transmission loss was computed and compared with experimental data at ten different operating conditions. The results were promising, but further developments were needed to use the method to predict the acoustic performance of turbines in operating points not already measured in experiments or in the turbine map. Galindo et al. [33] use a model proposed previously[91], which is similar to the one presented by Veloso et al.[147], for the main geometrical aspects of the turbine, while using simplified nozzles for the stator passages, conservation of rothalpy and mass flow for the rotor and a series of loss models. The results of the model are compared against a small set of experimental acoustic data, showing good agreement of the sound pressure level of the forward- and backward-travelling pressure waves in the frequency domain. The work presented in this chapter is based on the models documented by Galindo et al.[1], where their previous model[91] is adapted to use the work by Serrano et al. [30] to interpolate and extrapolate the turbine map in terms of adiabatic efficiency and mass flow. The frequency domain results showed good agreement against experimental data, with good capabilities to extrapolate to extreme off-design conditions.

In summary, a method to exploit the prediction capabilities of a one-dimensional variable geometry turbine model in order to produce acoustic transfer matrices for arbitrary operating conditions will be presented. These matrices can be computed during a preprocessing phase prior to further extensive computation campaigns. Coupled with quasi-steady calculations of the cycle-averaged turbine mass flow and efficiency, this method can be used to predict the turbine behaviour with an almost negligible computational cost. Other quasi-steady effects of the turbocharger can be easily taken into account, such as the compressor map extrapolation [31, 148], the heat transfer [27, 28, 29] or the mechanical losses in the bearings [32].

5.2   Method description and acoustic matrices generation

In this section, the method to generate and use a lookup table of acoustic transfer matrices is presented. Firstly, the experimental campaign to obtain steady-state and

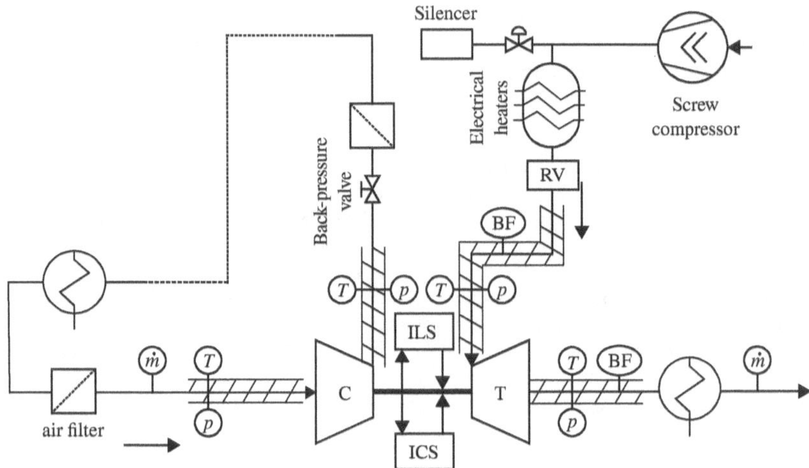

Figure 5.1: Gas stand schematic.

pulsating flow turbine results is described. Secondly, the basis of the one-dimensional simulations used during the generation of the lookup tables is shown, and the procedure for generating them from these simulations is presented. Finally, a simple process to compute an interpolated acoustic transfer matrix (IATM) is proposed.

5.2.1 Experimental campaign

The experimental campaign has been carried out in the CMT-Motores Térmicos gas stand for turbocharging research. This facility is specifically built for testing and measuring any type of automotive turbocharger under different conditions that has been presented in the previous chapters. The experimental method followed and the instrumentation used in order to obtain the experimental data for the validation here presented was described in detail in chapter 2. However, a basic sketch of the gas stand is again showed in Figure 5.1 and a summary of the procedure is here exposed for clarity's sake.

The gas stand uses a screw compressor to feed the turbine and its flow rate is controlled by means of its rotational speed and a discharge valve. The air passes through a set of electrical heaters to reach the desired temperature. When pulsating flow is required, a rotating valve is placed after the electrical heaters, generating the desired pulses.

The mass flow of the compressor and turbine were measured by means of hot film flow meters and temperature was measured by four type K thermocouples at the turbine inlet, turbine outlet, compressor inlet and compressor outlet. The average air pressure was measured by means of piezoresistive transmitters in the same stations as the flow temperature. The rotational speed of the turbocharger was measured by means of an induced currents transducer placed inside the compressor casing. Finally, an array of three piezoelectric transducers was placed at the turbine inlet and outlet during the pulsating flow characterisation campaign to get, via beamforming [101], the pressure pulses travelling downstream and upstream.

The experiments used in this work were done in quasi-adiabatic conditions, minimising the internal and external heat flows. The turbine was characterised first in steady-state conditions, obtaining its characteristic maps. Then, it was measured activating the rotating valve at its inlet, imposing pulsating flow. The fundamental frequency of the pulses generated were 50 Hz, 66 Hz and 100 Hz, corresponding to 1500 rpm, 2000 rpm and 3000 rpm in a four cylinders four stroke engine. Different turbocharger speeds and turbine stator vanes positions were evaluated, corresponding to different points in an equivalent engine. The experimental points are shown in Table 5.1.

Case label	Equivalent engine speed [rpm]	Approx. vanes position [%]	Fraction of maximum engine torque [%]
#1	2000	49	25 %
#2	2000	47	100 %
#3	3000	80	50 %
#4	3000	58	75 %
#5	3000	50	100 %
#6	3500	55	100 %

Table 5.1: List of cases of the single entry turbine pulsating flow experiments.

For the rest of the chapter, the pressure wave travelling from the rotary valve downstream towards the turbine inlet will be called incident pressure, whereas the wave travelling away from the turbine will be called reflected pressure. At the turbine outlet, the pressure wave travelling downstream will be called transmitted pressure. As the turbine outlet was not connected to an anechoic end, a second reflection pressure wave was also present, travelling towards the turbine at its outlet.

5.2.2 One-dimensional simulation

The simulation results that are presented in this study were obtained using OpenWAM [17], the 1D simulation integrated tool developed in CMT–Motores Térmicos, modified from its finite-differences formulation [94] to a finite-volume formulation. The method is described in detail in [1] and summarised in chapter 2.

The turbine is modelled as a combination of one-dimensional and zero-dimensional elements as seen in Figure 5.2. Station 0 is the domain inlet, upstream of the turbine. Station 1 is the turbine inlet. Station 2 is the volute lateral window. Station 2' is the stator nozzles throat section. Station 3 is the rotor inlet. Station 4 is the rotor outlet. Station 5 is the turbine outlet. Finally, station 6 is the domain outlet.

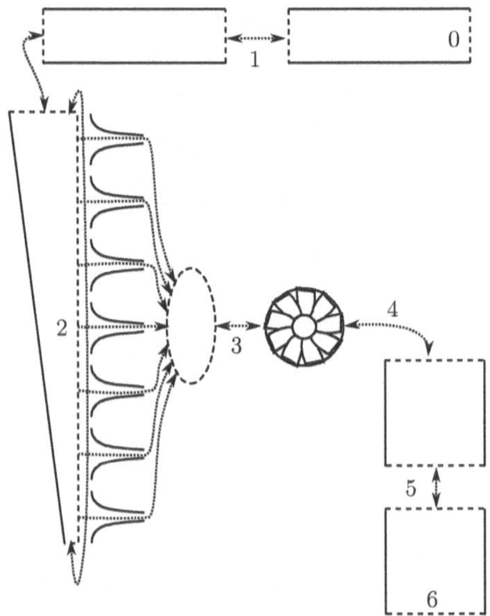

Figure 5.2: Turbine model schematic. The main sections are numbered.

The turbine mass flow capacity and its power output are taken into account by means of two variable-size nozzles, one for the stator and one for the rotor, and an energy sink in the volume between them. The amount of energy taken each iteration is equal to the turbine power output times the time step, which is computed from the turbine efficiency at these conditions. Meanwhile, the area of the stator and the rotor nozzles are computed from the reduced mass flow as described by Serrano et al. [25]. The turbine reduced mass flow and adiabatic efficiency as a function of its expansion ratio, reduced speed and vanes position can be obtained by means of the turbine map provided by the manufacturer. Although in some simple cases these maps can be used directly as interpolation lookup tables, normally an extrapolation model is needed. In this work, the model described by Serrano et al. [30] is used to extrapolate the map. The extrapolation performance of this model was validated against extreme off-design conditions [87]. The reduced mass flow $\dot{m}_{\text{red}}$ is computed as in Equation 5.1:

$$
\dot{m}_{\text{red}} = A_{\text{Neq}} \sqrt{\frac{\gamma}{\Re} \left(\frac{1}{\pi_{1\text{t},4\text{s}}} \right)^{\frac{1}{\gamma}}} \\
\cdot \sqrt{\frac{2}{\gamma - 1} \left[1 - \left(\frac{1}{\pi_{1\text{t},4\text{s}}} \right)^{\frac{\gamma-1}{\gamma}} \right]}
\tag{5.1}
$$

where A_{Neq} is the effective section of a nozzle that has the same reduced mass flow as the turbine with the same expansion ratio and working fluid, and $\pi_{1\text{t},4\text{s}}$ is the turbine total to static expansion ratio. The effective section is computed as in Equation 5.2:

$$
A_{\text{Neq}} = \frac{a \cdot A_4 \cdot \sqrt{1 + \dfrac{\sigma^2 \cdot \left[\left(\frac{D_4}{D_3} \right)^2 - 1 \right] + b}{\overline{\eta}}}}{\sqrt{1 + \left(c \cdot \frac{A_4}{A_{2'}} \right)^2 \cdot \dfrac{\left(\frac{1}{\pi_{2',4}} \right)^2}{\left\{ 1 - \eta \cdot \left[1 - \left(\frac{1}{\pi_{2',4}} \right)^{\frac{\gamma-1}{\gamma}} \right] \right\}^2}}}
\tag{5.2}
$$

where a, b and c are the fitting coefficients described by Serrano et al. [30]. η is the total-to-static turbine efficiency, $\overline{\eta}$ is the average turbine efficiency, D_3 is the rotor inlet diameter, D_4 is the rotor outlet diameter, $A_{2'}$ is the area at the stator nozzles outlet, A_4 is the rotor outlet area and $\pi_{2',4}$ is the expansion ratio of the rotor, computed as in Equation 5.3:

$$
\pi_{2',4} = 1 + d \cdot (1 - \pi_{\text{turb}})
\tag{5.3}
$$

where, again, d is a fitting coefficient. The turbine efficiency η can be computed as in Equation 5.4:

$$
\eta = -2 \cdot \left(\frac{\overline{D_4}}{D_3} \right)^2 \cdot \sigma^2 \\
+ \left(2 \cdot \frac{A_{\text{Neq}}}{A_1} \cdot z \cdot \tan \alpha_3 + \frac{\overline{D_4}}{D_3} \cdot \tan \beta_4 \right) \cdot \pi_{\text{turb}}^{-\frac{1}{\gamma}} \cdot \sigma
\tag{5.4}
$$

where $\overline{D_4}$ is the average rotor outlet diameter, σ is the blade speed ratio, α_3 is the flow angle at the rotor inlet, β_4 is the flow angle at the rotor outlet, and γ is the specific heat capacities ratio. z is a correction term that can be computed with the expression shown in Equation 5.5:

$$z = (e \cdot N_{\text{red}} + f) \cdot \sigma + \left(g \cdot N_{\text{red}} + h \cdot r^2 + i \cdot r + j\right) \tag{5.5}$$

where e, f, g, h, i, and j are fitting parameters, N_{red} is the turbine reduced speed, and r is the turbine stator vanes position.

Each time step, Equation 5.2 and Equation 5.4 are solved iteratively until convergence. Then, the effective area is used to compute the effective section of the stator nozzles and the rotor and set as the area of the boundary conditions of the one-dimensional and zero-dimensional elements in contact with stations 2 and 4. The efficiency is used as described previously.

In the next section, the set of simulations carried out using this model needed to obtain the acoustic matrices will be described.

5.2.3   Acoustic transfer matrix generation

Two transient simulations are performed for each turbine map point where the acoustic transfer matrix is generated. In these simulations, the turbine inlet and outlet are connected to incident pressure boundary conditions, where the pressure waves travelling towards the turbine are imposed as a time-varying signal. Also, the rotational speed of the rotor is kept constant and equal to the one corresponding at that turbine map point. During the first simulation, the pressure wave at the turbine outlet has a constant value whereas the pressure wave at the turbine inlet introduces low amplitude white noise with an average pressure value corresponding to the turbine map point. In the second simulation, the turbine inlet pressure wave is kept constant while low amplitude white noise is imposed at the turbine outlet.

During the first simulation, the turbine outlet acts as an anechoic end, so the pressure waves travelling upstream at the turbine inlet and downstream at the turbine outlet are produced only due to the turbine inlet imposed pressure wave. During the second simulation, the roles are inverted and the backward-travelling pressure pulse at the turbine inlet and the forward-travelling pressure pulse at the turbine outlet are a consequence of the imposed pulse at the turbine outlet boundary condition. The results of the simulation are processed, discarding the first $0.25\,\text{s}$ to ensure convergence, approximating the power spectral density of the pressure waves by means of Welch's method [102] with a Hanning window. Then, an expression in the form of Equation 5.6 can be computed.

$$\begin{pmatrix} R_{\text{in}} & T_{\text{out,in}} \\ T_{\text{in,out}} & R_{\text{out}} \end{pmatrix} \cdot \begin{pmatrix} P_{\text{in}}^{+} \\ P_{\text{out}}^{-} \end{pmatrix} = \begin{pmatrix} P_{\text{in}}^{-} \\ P_{\text{out}}^{+} \end{pmatrix} \tag{5.6}$$

In the left term of Equation 5.6, the matrix corresponds to the acoustic transfer matrix (ATM). In the same equation, the subscript *in* represents the turbine inlet and the subscript *out* the turbine outlet. The pressures P are computed in the frequency domain. The superscript $^+$ refers to a pressure wave travelling downstream, whereas $^-$ refers to a pressure wave that travels upstream. The reflection coefficients R_{in} and R_{out} can be computed as in Equation 5.7.

$$R_{\text{in}} = \left.\frac{P_{\text{in}}^-}{P_{\text{in}}^+}\right|_{\text{first sim.}} \quad , \quad R_{\text{out}} = \left.\frac{P_{\text{out}}^+}{P_{\text{out}}^-}\right|_{\text{second sim.}} \tag{5.7}$$

The transmission coefficients $T_{\text{out,in}}$ and $T_{\text{in,out}}$ are computed as in Equation 5.8:

$$T_{\text{out,in}} = \left.\frac{P_{\text{in}}^-}{P_{\text{out}}^-}\right|_{\text{second sim.}} \quad , \quad T_{\text{in,out}} = \left.\frac{P_{\text{out}}^+}{P_{\text{in}}^+}\right|_{\text{first sim.}} \tag{5.8}$$

This procedure is shown schematically in Figure 5.3. Once the acoustic transfer matrices are assembled, they supplant the whole turbine model represented in Figure 5.2 in terms of fluid acoustics computation.

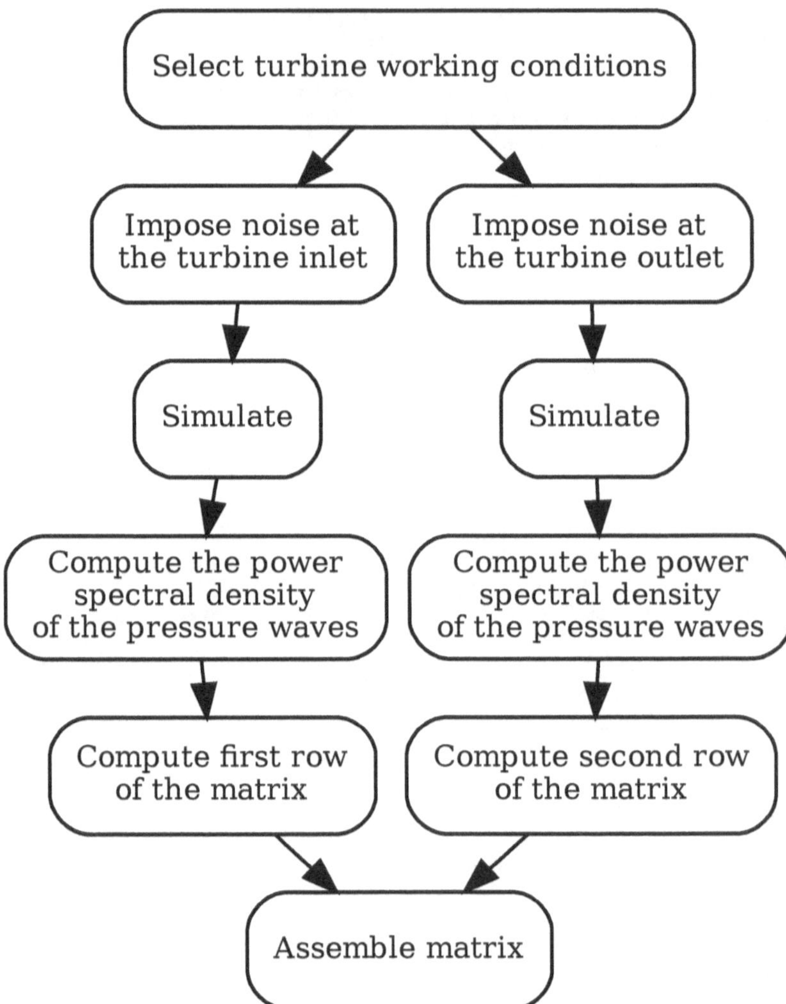

Figure 5.3: Flowchart of the acoustric transfer matrix generation.

5.2.4   Acoustic transfer matrix interpolation from lookup table

After generating the lookup table, a standard interpolation method can be used to obtain the acoustic transfer matrix function needed for any turbine operating point. For a point of reduced speed N_{red}, expansion ratio π_{turb} and vanes position r, six values of the acoustic transfer matrix lookup table are searched for: the immediately lower and higher reduced speeds, $N_{\text{red}}^{\text{lower}}$ and $N_{\text{red}}^{\text{upper}}$; the immediately lower and higher expansion ratios, $\pi_{\text{turb}}^{\text{lower}}$ and $\pi_{\text{turb}}^{\text{upper}}$; the immediately lower and higher vanes positions, r^{lower} and r^{upper}. Then, a simple trilinear interpolation can be performed, using the values of $T_{\text{in,out}}$, $T_{\text{out,in}}$, R_{in} and R_{out} stored in the lookup table for the eight combinations of lower and higher values of reduced speed, expansion ratio and vanes position.

The mass flow and efficiency of the turbine can be computed directly from Equation 5.1 and Equation 5.4, and the flow speed and temperature can be easily obtained and used to get the behaviour of the elements downstream of the turbine.

5.3   Model validation and sensitivity study

The objective of this section is to evaluate the performance of the IATM method in terms of turbine acoustic prediction when compared to conventional one-dimensional simulations as well as to demonstrate its potential and simple implementation. With this purpose, an experimental validation using the data described in chapter 2 and documented in [1] is first presented. Secondly, a sensitivity study in terms of interpolation mesh resolution is performed to explore the strengths and limitation of the method. Finally, a complete analysis of the turbine acoustics with the main operation variables of the turbine is exposed.

5.3.1   Experimental validation

The experimental turbine map shown in Figure 5.4 is fed to the model and, during the preprocessing step, a complete extrapolated map based on equivalent effective areas and efficiencies[30] is generated.

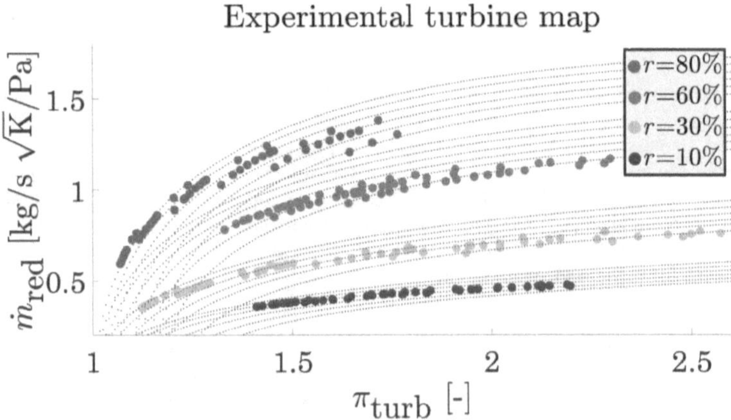

Figure 5.4: Turbine map obtained from steady state measurements.

The 1D model used as a base for the interpolation method has been previously validated (chapter 2). However, the next results include the simulations obtained using the IATM method, with the aim of performing a comparison with the fully 1D results. The IATM results are obtained after the interpolation of the acoustic matrices calculated from the 1D model following the method described in the section called Method description and acoustic matrices generation. Thus, when comparing with the experiments, any improvement of the IATM in the accuracy should be interpreted as a matter of chance. As explained in chapter 2, for carrying out the validation, the incident and second reflection components are imposed in the inlet and outlet of the turbine respectively. The reflected and transmitted waves are predicted and compared with the experiments to evaluate the capability of the model to recreate the turbine acoustics. Considering that the simulation cost of the IATM method would be significantly lower when only a pre-processing is needed, the usefulness of the tool will be evident as long as the results are reasonably similar to the ones provided by the 1D model.

From Figure 5.5 to Figure 5.10 the results obtained with the 1D model and with the IATM are presented, along with the experimental results. A summary of the conditions of the main experimental points is presented in Table 5.1, covering different operation conditions in terms of engine speed, engine load, and variable geometry turbine (VGT) position. The plots are divided into two subplots. In the first subplot, the power spectral density (PSD) is showed from 0 Hz to 650 Hz. On the other hand, the subplot on the bottom indicates the error against experiments of the harmonics, calculated for both the 1D and the IATM method.

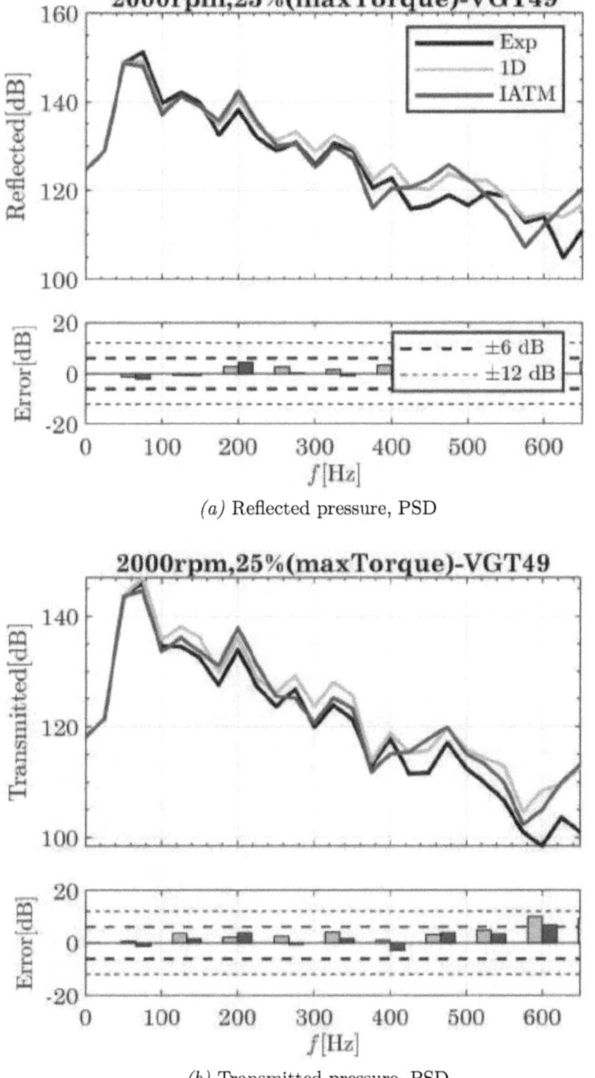

(a) Reflected pressure, PSD

(b) Transmitted pressure, PSD

Figure 5.5: PSD results validation. Reflected and transmitted for case 1(a, b).

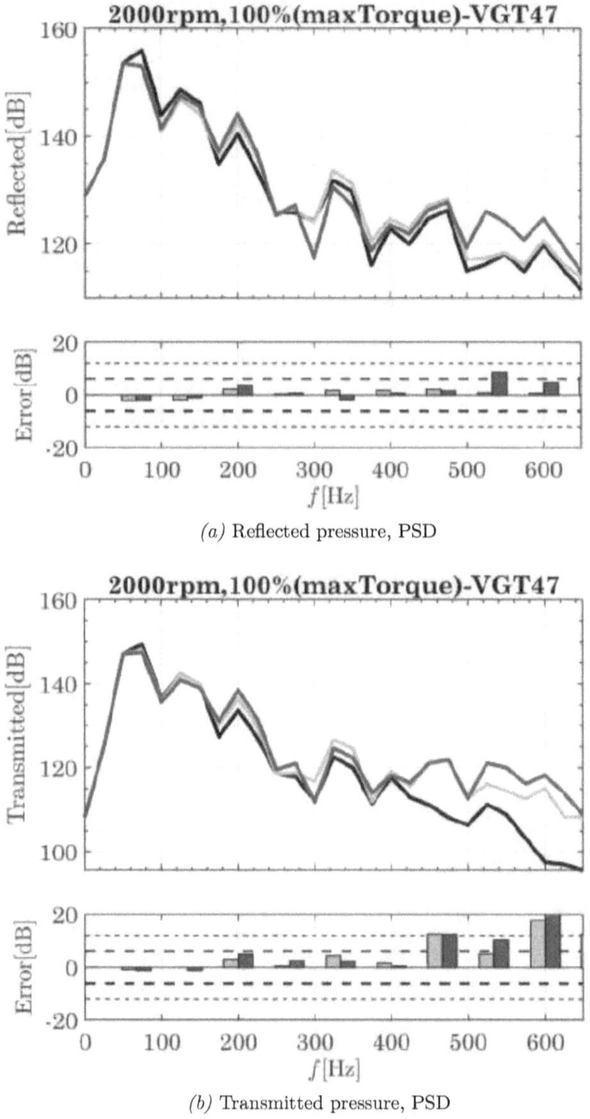

(a) Reflected pressure, PSD

(b) Transmitted pressure, PSD

Figure 5.6: PSD results validation. Reflected and transmitted for case 2(a, b).

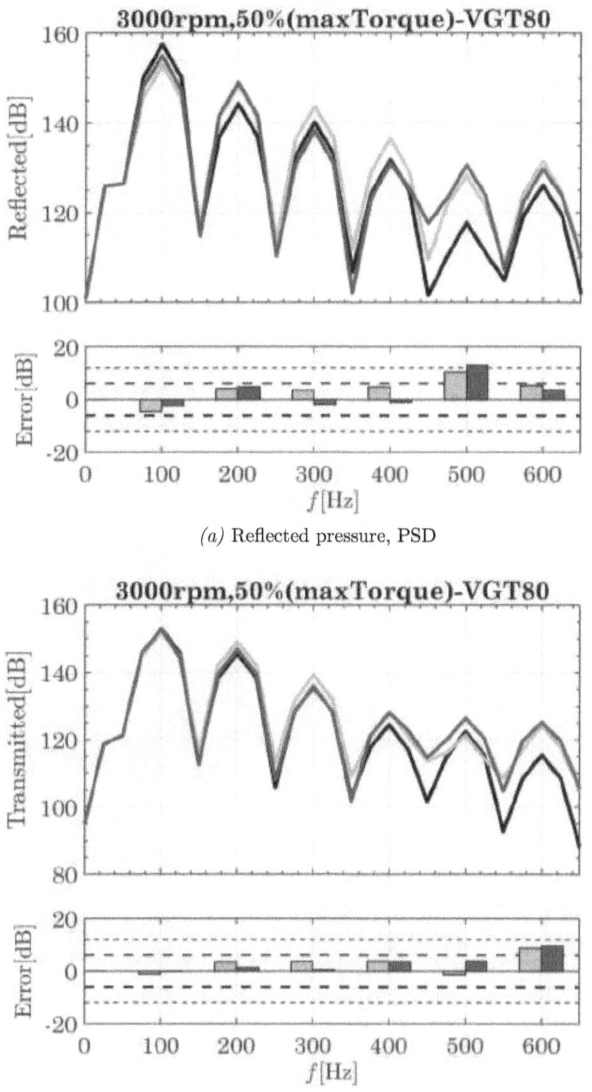

(a) Reflected pressure, PSD

(b) Transmitted pressure, PSD

Figure 5.7: PSD results validation. Reflected and transmitted for case 3(a, b).

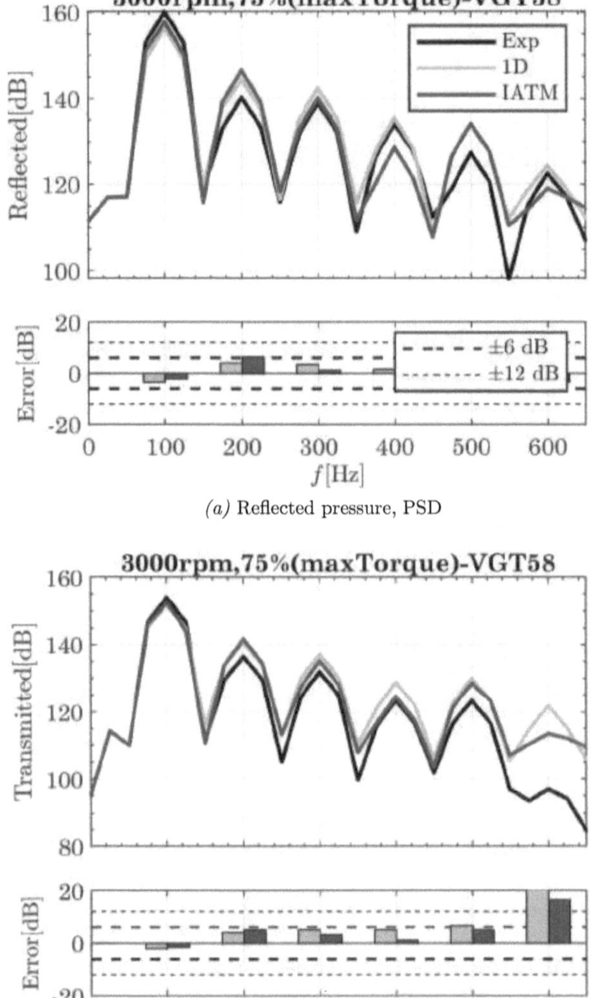

(a) Reflected pressure, PSD

(b) Transmitted pressure, PSD

Figure 5.8: PSD results validation. Reflected and transmitted for case 4(a, b).

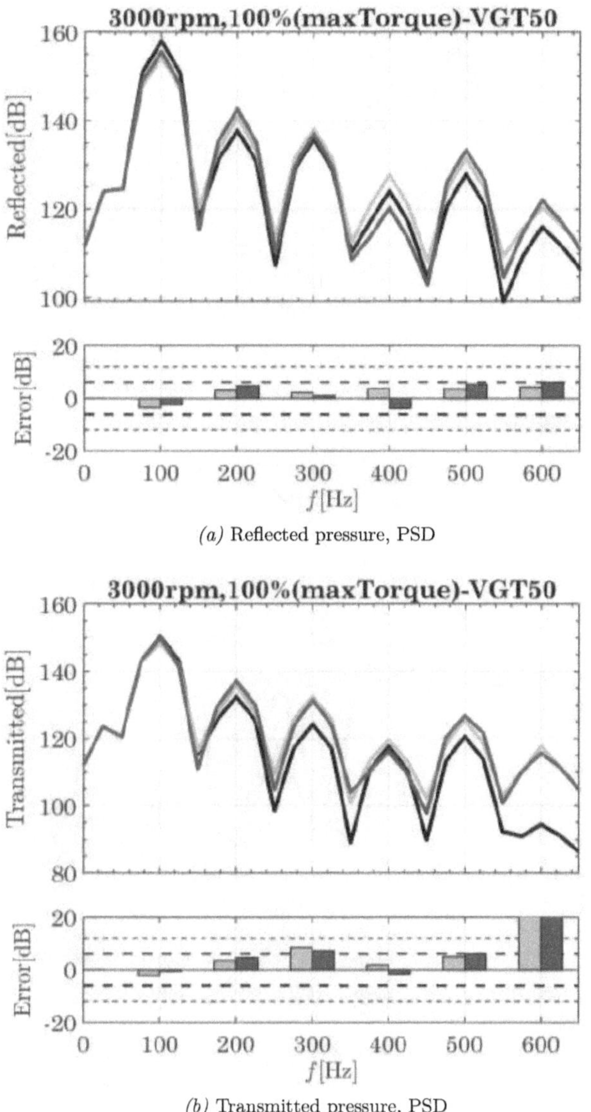

(a) Reflected pressure, PSD

(b) Transmitted pressure, PSD

Figure 5.9: PSD results validation. Reflected and transmitted for case 5(a, b).

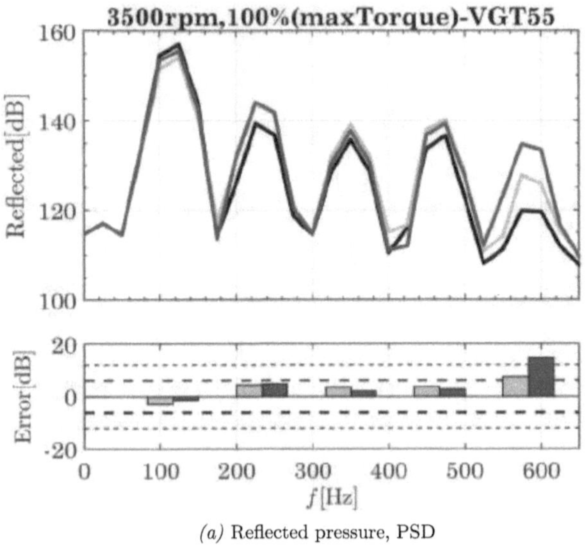

(a) Reflected pressure, PSD

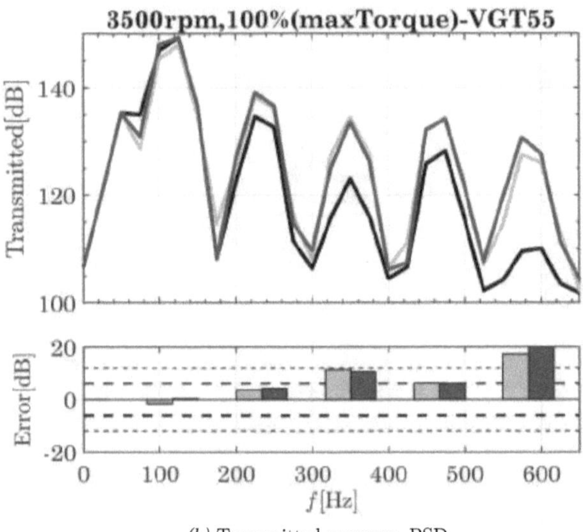

(b) Transmitted pressure, PSD

Figure 5.10: PSD results validation. Reflected and transmitted for case 6(a, b).

Observing the results from Figure 5.5 to Figure 5.10 it can be stated that the IATM method effectively reproduced the 1D performance. Not in vain, both the PSD graph and the error are almost identical using both models, for all the cases, in both reflected and transmitted waves. The biggest difference between IATM and 1D is seen probably in the reflection of case 6 at 600 Hz. When comparing with experiments, the error is maintained lower than 6 dB for frequencies below 500 Hz in almost all the cases,with the exception of case 2-transmitted (5.6(b)), case 3-reflected (5.7(a)) and case 6-transmitted (5.10(b)), where an error of around 12 dB appears near 500 Hz.

As it has been previously stated, the aim of this validation is not to evaluate the discrepancies with the experiments but to prove that the IATM method provides similar results as the considerably more expensive 1D model. In this sense, in the light of the results presented, the IATM method arises as a reliable and computationally efficient alternative to the 1D method for turbine acoustics prediction in low and medium frequencies.

5.3.2 Lookup table resolution sensitivity study

The procedure that has been followed in order to build the acoustic transfer matrices have been previously exposed in section 5.2, along with the description of the interpolation method from lookup tables. In the following paragraphs, a study of the resolution of the interpolation mesh, in addition to some indications of the criteria for selecting the mesh points, are presented. This information could be relevant, considering that different mesh resolutions will lead to different computational time during the computation of the interpolation matrices. Furthermore, the specific selection of the points chosen for building the mesh might be a good indicator of the robustness of the procedure, as the benefit of the method relies on the ability to predict operating points even when they are reasonably far from the ones that constitute the interpolation mesh.

Configuration	Mesh re-solution	N_{red} $[\mathrm{Hz\,K^{-0.5}}]$	r [%]	π_{turb}
Exp	-	50, 95, 85, 104, 106, 107	49, 47, 80, 58, 50, 55	1.35, 2.34 1.66, 2.28, 2.58, 2.4
Base mesh	108	43, 86, 129	10, 40, 90	1.35, 2.34 1.66, 2.28, 2.58, 2.4
Coarse Mesh	54	43, 86, 129	10, 40, 90	1.1, 1.6, 2.6
Thin mesh	216	43, 86, 129	10, 30, 60, 90	1.1, 1.2, 1.4, 1.6, 1.8, 2, 2.2, 2.4, 2.6

Table 5.2: Configuration details of the mesh sensitivity study.

In Table 5.2 there is a summary of the relevant information used to perform this study. The first row includes the data of the main turbine parameters for the six experimental points used in the previously discussed experimental validation. The second row, referenced as *base mesh*, includes the data of the main turbine parameters used for the mesh configuration selected for the experimental validation. Two additional rows are included, one for a *thin mesh* configuration and another for a *coarse mesh*. These two additional computational meshes have been tested for performing a mesh sensitivity analysis. For each additional mesh configuration, Table 5.2 indicates the turbine operation parameters used for computing the interpolation mesh. As the reader can notice, the operating parameters selected for building the mesh are different from the ones in the experiments in all three configurations, otherwise it would no be possible to test the reliability of the interpolation.

The IATM method has been applied using the additional two mesh configurations, *coarse mesh* and *thin mesh*, and then compared with experiments and the 1D model. For clarity's sake, only a summary of the mesh sensitivity study results will be shown, as the conclusion obtained is sufficiently evident to be presented in a single chart. In Figure 5.11, the root-mean-square error obtained for the reflected and transmitted component, is presented for all the mesh configurations in the six cases studied. The RMSE is computed using the error of the first six harmonics when comparing the experiment and the simulations of each of the IATM mesh configurations.

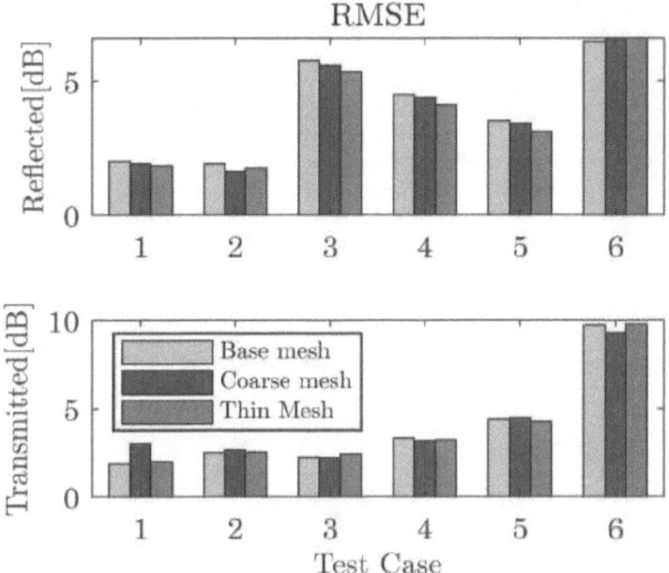

Figure 5.11: RMSE of the IATM with different mesh resolutions.

As it can be observed in Figure 5.11, results of RMSE are extremely similar for all the mesh configurations in both the reflected and the transmitted components. Furthermore, the error is in general less than 5 dB, except for case 6.

From these results it can be concluded that the interpolation method is robust, stable and accurate, as it provides very similar results to the 1D model, even with the *coarse mesh* where only 27 points are used for the mesh computation.

5.3.3   Computational cost

In order to obtain the acoustic transfer matrix of one single operating point, two simulations of 1.25 s are performed and their results are processed. Running sequentially, the total time in an Intel ®Xeon ®E5-2640 v4 CPU at 3 GHz is 90 s.

The computational cost for computing the matrices increases linearly with the density of the mesh as expected. In particular, the coarse mesh takes half of the time as the base mesh, and the thin mesh takes double time of the base mesh, as it has the double quantity of points. However, the interpolation matrices are computed during preprocessing and thus, the computational cost might not be determinant when configuring the method. The relevant fact is that using this method, the turbine acoustics are already obtained before the simulation and thus, only lookup table computations are required during further uses. Also, the procedure of filling the lookup table is extremely parallel, so it can be naturally run in multiple computational cores. In fact, once the calculation of the acoustic matrices is amortised in the preprocessing, faster than real time conditions can be easily reached. In contrast, the Intel machine above mentioned needs around 15 seconds to compute each second of the same simulation case when running the complete 1D model represented in Figure 5.2.

A faster than real time turbine model with acoustic capabilities such as the one presented in this work can be easily coupled with other real time or faster than real time simulation elements, including some of the newest models for catalytic converters [149] or gas exchange models [150].

5.3.4 Turbine acoustic performance analysis

At this stage, it has been demonstrated that the IATM method provides very similar results as a 1D model in terms of accuracy, but implying a much lower computational cost. The results of the interpolation mesh is, furthermore, considerably robust even when using a low quantity of mesh points (coarse mesh). But it is not less important to prove the usefulness of the method with an example of a turbine acoustic performance analysis. The next results use the same turbine as the one previously presented. From the steady state experimental test, the turbine map of Figure 5.4 is obtained. With this turbine map, the interpolation mesh is generated and then, a complete analysis of the turbine acoustics can be performed using the IATM method.

For the following analysis, the base mesh configuration is used. The equivalent effective area of the 54 turbine operating points of the interpolation mesh is shown in Figure 5.12. The effective area will have a crucial role in the acoustics performance as it will be shown in the following paragraphs.

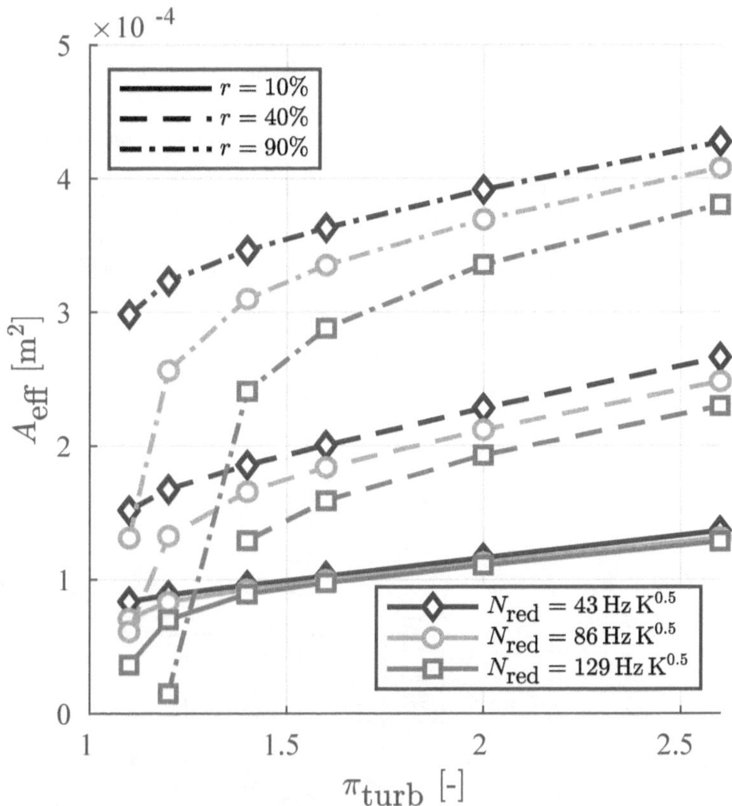

Figure 5.12: Equivalent turbine effective area of the operating points used for the generation of the interpolation mesh (being r the vane position).

Once the preprocessing is finished and the acoustic matrices are obtained, there are several possibilities in terms of ways of studying the acoustic performance of the turbine. However, the study presented is focused in the behaviour of the transmission loss (TL) when confronted to the main parameters of the turbine operation, i.e., the expansion ratio, the vanes position and the turbocharger speed, as well as the equivalent effective area shown in Figure 5.12. The transmission loss is defined as the ratio between the transmitted pressure wave and the incident pressure wave, $P_{\mathrm{out}}^+/P_{\mathrm{in}}^+$, and is expressed in dB. Thus, it quantifies the attenuation that the incident wave faces when travelling through the turbine towards the outlet pipe (inlet of the aftertreatment system). The main objective of the following results is to illustrate the potential of the method when analysing the acoustic performance rather than to fully understand the specific behaviour of this turbine specimen.

In Figure 5.13, Figure 5.14 and Figure 5.15, the evolution of the transmission loss against frequency is shown for a set of turbine operating points at different stator vanes positions and reduced speeds. Result from Figure 5.13 corresponds to a turbine point with a vanes position 10 % and a reduced speed of $129\,\mathrm{Hz\,K^{-0.5}}$, Figure 5.14 corresponds to a point of 40 % and a reduced speed of $86\,\mathrm{Hz\,K^{-0.5}}$ and the result from Figure 5.15 corresponds to a vanes position of 90 % and a reduced speed of $129\,\mathrm{Hz\,K^{-0.5}}$.

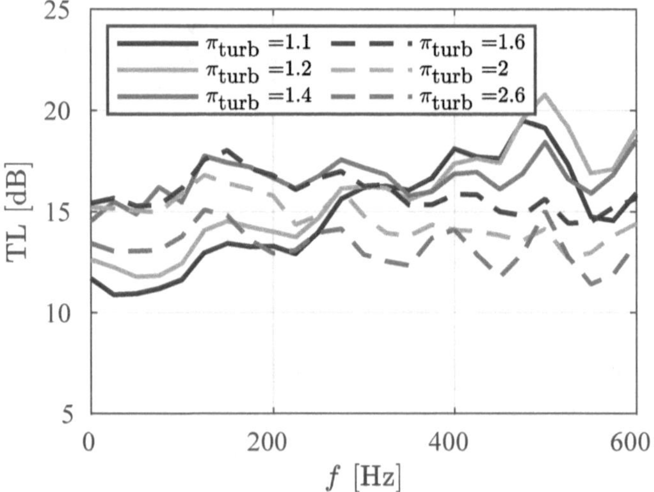

Figure 5.13: Vanes position at 10 %, reduced speed equal to $129\,\mathrm{Hz\,K^{-0.5}}$

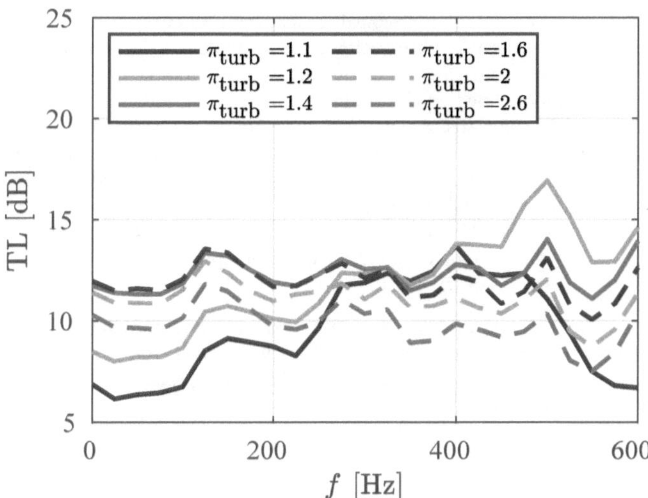

Figure 5.14: Vanes position at 40 %, reduced speed equal to $86\,\mathrm{Hz\,K^{-0.5}}$

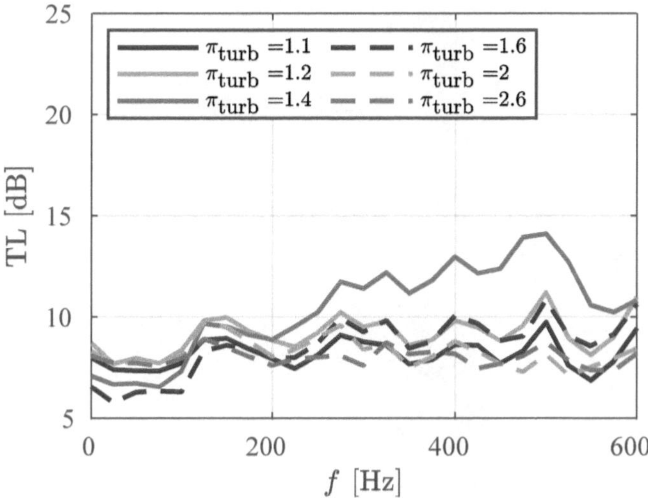

Figure 5.15: Vanes position at 90 %, reduced speed equal to $129\,\mathrm{Hz\,K^{-0.5}}$

In relation to the above results regarding the evolution of the transmission loss with the frequency, some interesting observations can be made. Firstly, the transmission loss decreases as the stator vanes open. The flow restriction produced by the stator is lower when the vanes are more open, which could explain this effect. Secondly, in general, and for a given frequency, the transmission loss increases first as the pressure ratio increases, reaches a maximum value and then decreases again. This value where the losses are maximum probably coincides with critical or close to critical conditions in the flow, which would also indirectly affect the efficiency. Finally, the pressure ratio at which the maximum value is presented is close to the point where the slope of the effective section curve of Figure 5.12 starts to be constant. The specific pressure ratio at which this effect happens is also a function of the frequency.

In Figure 5.16 the transmission loss averaged at different frequency bands is confronted to the different expansion ratios , where the trends shown from Figure 5.13 to Figure 5.15 are more easily identifiable. Figure 5.17 shows the same results, but plotted against the turbine effective area.

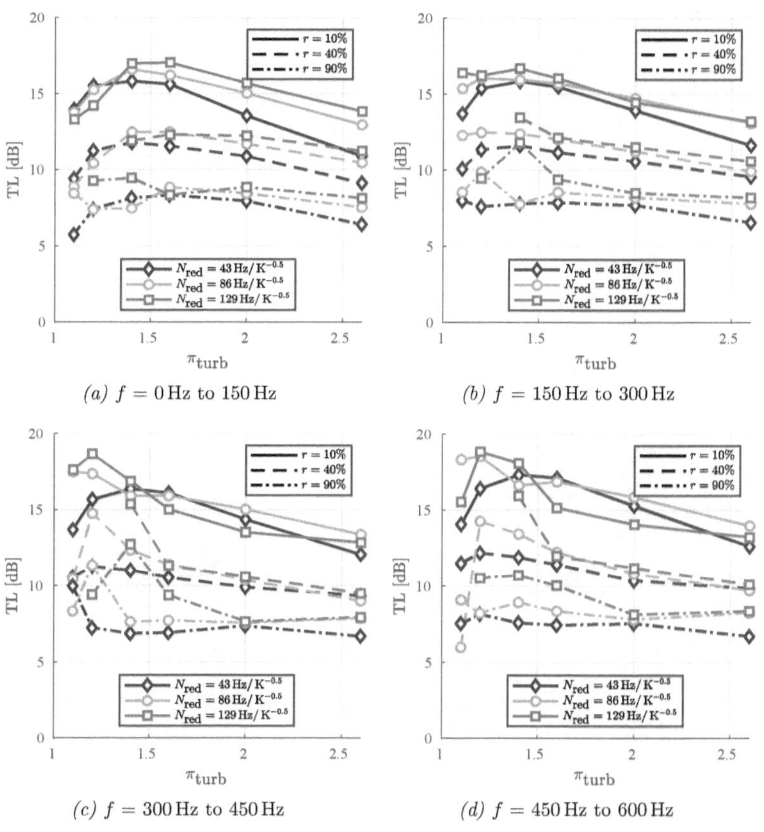

(a) $f = 0\,\mathrm{Hz}$ to $150\,\mathrm{Hz}$

(b) $f = 150\,\mathrm{Hz}$ to $300\,\mathrm{Hz}$

(c) $f = 300\,\mathrm{Hz}$ to $450\,\mathrm{Hz}$

(d) $f = 450\,\mathrm{Hz}$ to $600\,\mathrm{Hz}$

Figure 5.16: Transmission loss against the expansion ratio of the turbine, averaged at different frequency bands.

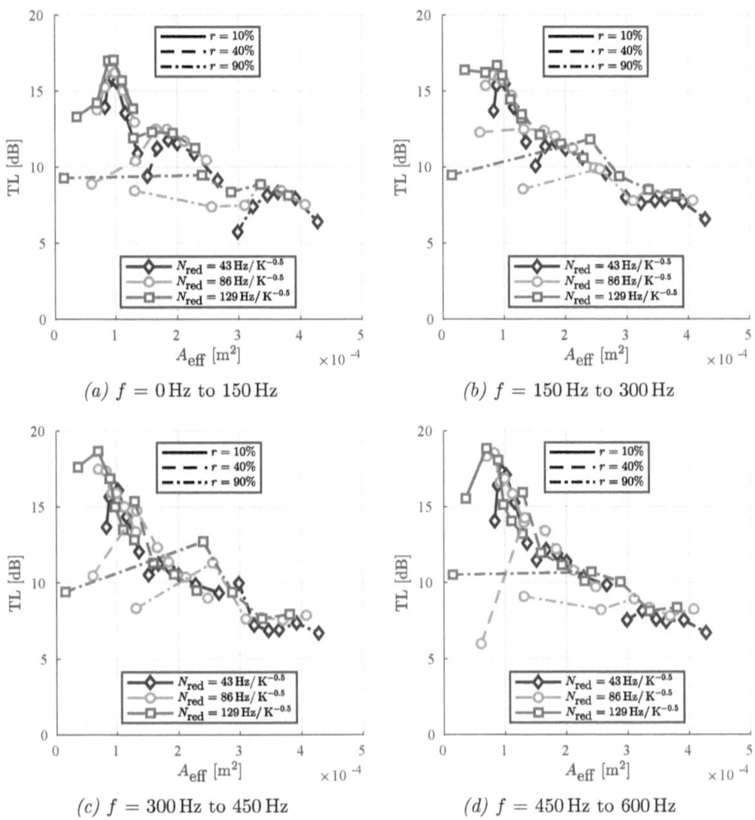

(a) $f = 0\,\mathrm{Hz}$ to $150\,\mathrm{Hz}$ *(b) $f = 150\,\mathrm{Hz}$ to $300\,\mathrm{Hz}$*

(c) $f = 300\,\mathrm{Hz}$ to $450\,\mathrm{Hz}$ *(d) $f = 450\,\mathrm{Hz}$ to $600\,\mathrm{Hz}$*

Figure 5.17: Transmission loss against the effective area of the turbine, averaged at different frequency bands.

From Figure 5.16 it can be observed how, again, the transmission losses reach a maximum value at quite low expansion ratios. The main dependency is with the opening of the stator vanes, whereas the reduced speed effects are more noticeable at low expansion ratios. After the maximum transmission loss is reached, there is a small drop in transmission losses with growing frequencies. Furthermore, from Figure 5.17 it can be stated that the transmission losses grow as the effective area drops. However, the relationship between the effective area and the transmission losses is not bijective: other effects seems to be affecting the results. For a given effective area, the turbine efficiency might be different depending on the position of its vanes and its reduced speed. These changes in turbine adiabatic efficiency explain the rest of the changes in the turbine acoustic behaviour.

Figure 5.18 shows the evolution of the turbine adiabatic efficiency with the expansion ratio and the effective area. The transmission losses deviate more from a single bijective function of the effective area in points where the adiabatic efficiency is very far from design conditions, at very high blade speed ratios or low expansion ratios. In fact, the efficiency reaches lower-than-zero values in the extreme off-design conditions, although the model limits this value to -1. From this, it is clear that it is necessary a model able to extrapolate accurately not only the mass flow but also the turbine adiabatic efficiency at far off-design conditions where the turbine consumes power as a brake. These conditions are taken into account in the model used in this work[30] or, more recently, the losses model by Serrano et al.[98]. Also, specific experiments are needed to validate the results at these conditions.

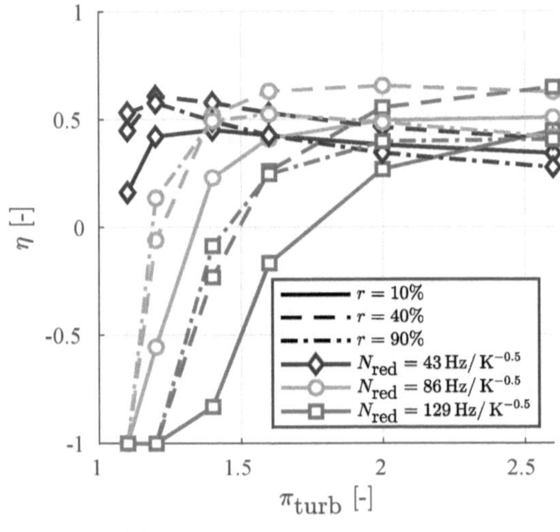

(a) Efficiency versus expansion ratio

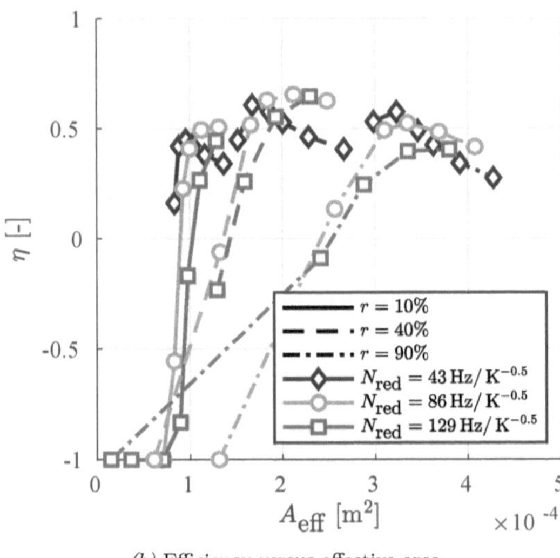

(b) Efficiency versus effective area

Figure 5.18: Turbine efficiency against expansion ratio (a) and effective section(b).

As a conclusion, the position of the stator vanes, mainly affecting the effective area, is the parameter that most affects the TL. The expansion ratio and reduced speed also change the effective area, although to a lesser extent. Also, the behaviour of the turbine is not exactly that of a simple connection between two one-dimensional elements of different section, as the effect of its efficiency is also visible in the results.

5.4 Model adaptation for twin-entry turbine

In the previous sections of this chapter, a fast method to predict the acoustics of a single entry turbines has been explained in detail and a complete validation has been exposed. In the present section, a first procedure to adapt the IATM method to twin-entry turbines is proposed and some results are exposed. Nevertheless, the goal of this section is not to perform an exhaustive validation, instead, the intention is to assess the problem and obtain some preliminary results in order to identify the limitations of the method that will have to be confronted in further studies.

5.4.1 Method description

In order to apply the IATM method to the twin-entry turbine, there are two main problems to confront. Firstly, it is necessary to have a reliable twin-entry turbine one-dimensional model in order to perform the simulations from which the interpolation lookup tables are generated. Secondly, the method to compute the acoustic transfer matrix needs to be adapted in order to take into consideration the additional turbine branch.

The set of simulations have been obtained again using OpenWAM as in the previously described single entry turbine application of the method. As it has been exposed in the validation campaigns of the model (section 4.3), it is possible to correctly reproduce the pulsating performance of a twin-entry turbine in a wide range of the frequency spectra. In order to firstly evaluate the goodness of the acoustic transfer interpolation procedure, it will be adequate to compare the output provided by the classic one-dimensional model, with the output provided using the interpolation. As the one-dimensional model has been described in detail in previous chapters of this book, the following paragraphs are focused in the generation of the acoustic transfer matrices.

As commented before, the twin-entry turbine introduces an additional focus of oscillations that corresponds to the additional inlet branch. Both turbine inlets and the turbine outlet are connected to incident pressure boundary conditions. Three simulations must be carried out for each of the turbine map points in which the transfer matrix is to be generated. With a similar philosophy as in the single entry turbine transfer matrix generation, in each simulation, a specific pulse is imposed in one of the boundaries, while the other two are kept constant. Thus, for each simulation, the two not pulsating extremes are actuating as anechoic ends. Again, results of the simulations are processed approximating the power spectral density of the pressure waves using Welch's method with a Hanning window. Now, as seen in Equation 5.9 the expression of the **acoustic** transfer matrix has an additional dimension:

$$
\begin{pmatrix} R_{31} & T_{32,31} & T_{4,31} \\ T_{31,32} & R_{32} & T_{4,32} \\ T_{31,4} & T_{32,4} & R_4 \end{pmatrix} \cdot \begin{pmatrix} P_{31}^+ \\ P_{32}^+ \\ P_4^- \end{pmatrix} = \begin{pmatrix} P_{31}^- \\ P_{32}^- \\ P_4^+ \end{pmatrix}
\tag{5.9}
$$

In the left term of Equation 5.9, the matrix corresponds to the acoustic transfer matrix (ATM). In this case, instead of *in* and *out*, new subscripts are used to differentiate between turbine branches for clarity's sake. This way, *31* and *32* subscripts indicate the turbine inlets for branch 1 and 2 respectively, whereas the subscript *4* references the turbine outlet. The pressures P are computed in the frequency domain. The superscript $^+$ refers to a pressure wave travelling downstream, whereas $^-$ refers to a pressure wave that travels upstream.

The reflection and transmission coefficients are obtained with the following procedure:

$$
R_{31} = \frac{P_{31}^-}{P_{31}^+}, \quad T_{31,32} = \frac{P_{32}^-}{P_{31}^+}, \quad \text{and} \quad T_{31,4} = \frac{P_4^+}{P_{31}^+}
\tag{5.10}
$$

are obtained maintaining the turbine inlet *2* and the turbine outlet as anechoic ends;

$$
R_{32} = \frac{P_{32}^-}{P_{32}^+}, \quad T_{32,31} = \frac{P_{31}^+}{P_{32}^+}, \quad \text{and} \quad T_{32,4} = \frac{P_4^+}{P_{32}^+}
\tag{5.11}
$$

are obtained maintaining the turbine inlet *1* and the turbine outlet as anechoic ends, and

$$
R_4 = \frac{P_4^+}{P_4^-}, \quad T_{4,31} = \frac{P_{31}^-}{P_4^-}, \quad \text{and} \quad T_{4,32} = \frac{P_{32}^-}{P_4^-}
\tag{5.12}
$$

are obtained simulating both turbine inlets as anechoic ends and imposing the pulse in the turbine outlet boundary.

5.4.2 Preliminary results

Before validating the performance of the IATM method when predicting the twin-turbine acoustics, it is important to analyse the turbine maps in order to anticipate the limitations and non-linearities that could appear in the turbine behaviour. The equivalent turbine effective area for shroud and hub was obtained compared to the expansion ratio for different MFR and turbocharger speeds. The equivalent effective area versus expansion ratio for different MFR and 150 krpm is shown for the shroud branch Figure 5.19 and hub branch Figure 5.20. The evolution of the effective area presents clear similarities when comparing it with the equivalent effective area of a VGT. In the same way that higher openings in the vane position lead to higher values of effective area, higher values of MFR will translate into higher values of effective area in the case of the shroud and lower values of effective area for the hub. Focusing the attention in a given MFR and branch, the trend followed by the effective area is also similar to the previously presented for the single entry turbine. The trend is clearly linear for the medium to higher range of expansion ratio, whereas for the lower range it appears to be more exponential, which will introduce limitations when trying to linearly characterise the turbine.

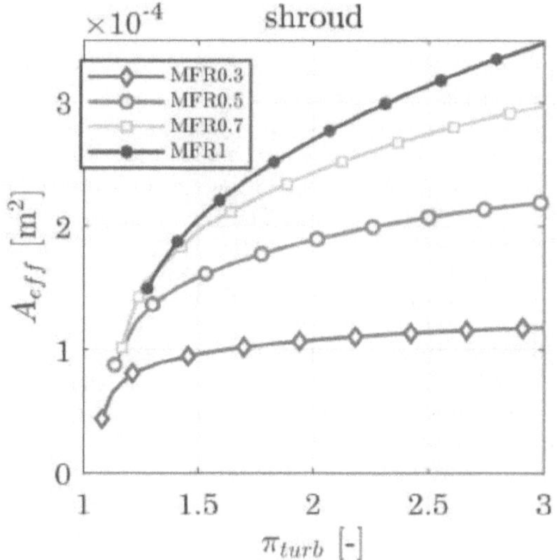

Figure 5.19: Equivalent turbine effective area versus expansion ratio in shroud branch.

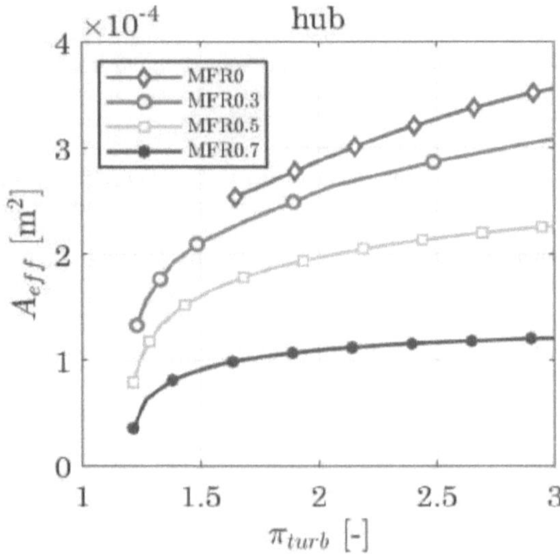

Figure 5.20: Equivalent turbine effective area versus expansion ratio in hub branch.

Following the procedure described in subsection 5.4.1, it is possible to compute the acoustic transfer matrices and characterise the turbine acoustics for any set of operating conditions. The following analysis and set of results are obtained making use of the experimental campaign from chapter 3. The fact that there are two inlet turbine branches introduces some difficulties and limitations in the interpolation procedure due to the fact that during the computation it is necessary to go over a set of mass flow ratios in addition to a set of expansion ratios. Because of that, it is convenient to analyse first the acoustic transfer matrices without interpolating, i.e., using the precise averaged conditions of the experimental points. For this purpose, the mean pressure and temperature from the experiments are imposed in each of the boundaries, and, imposing turbocharger speed as well. A random pressure pulse with an amplitude between 0 and 10 kPa is imposed sequentially in one branch while maintaining the other two branches as anechoic. With this approach, it is possible to discern whether the acoustic response of the turbine is diverse when imposing similar conditions but in different inlet branches, even with a low amplitude pulse that pertains to the linear domain.

To assess this matter, Figure 5.21 and Figure 5.22 show the components of the acoustic transfer matrix corresponding to similar operating conditions but pulsating in shroud and hub branches respectively. In particular, the conditions are an MFR of 0.5, a turbocharger speed of 97 krpm and an equivalent engine speed of 4600 rpm. From these figures, observing the behaviour of the components, it is clear that the geometry of the entries and volutes of the turbine introduce difference in the response even for small amplitudes, which is coherent with the results exposed in subsection 3.3.3. The 1D model used as a base for generating the acoustic matrices had already proved to be able to reproduce these asymmetries between branches (see section 4.3) for non-linear pulsating conditions. However, from these results, it can be inferred that the model appears to also capture the different acoustic response even when very low amplitude pulses are imposed.

After obtaining the acoustic matrices for each of the experimental points, the natural next step is to perform a first comparison between the pressure decomposition results from the experiments, and the equivalent results approximated by the acoustic transfer matrix computed in each of these points. For the following pressure decomposition results in the frequency domain, the nomenclature introduced in subsection 5.4.1 is particularized for the specific twin-turbine of the experiments. This way, considering shroud and hub branches referenced as 31 and 32 respectively, if the graph shows the pulsating in shroud branch results, then the subscript 31^- indicates the reflected component, whereas the subscript 32^- indicates the indirect transmitted. On the contrary, if pulses are performed in the hub, 32^- indicates reflected and 31^- indicates indirect transmitted.

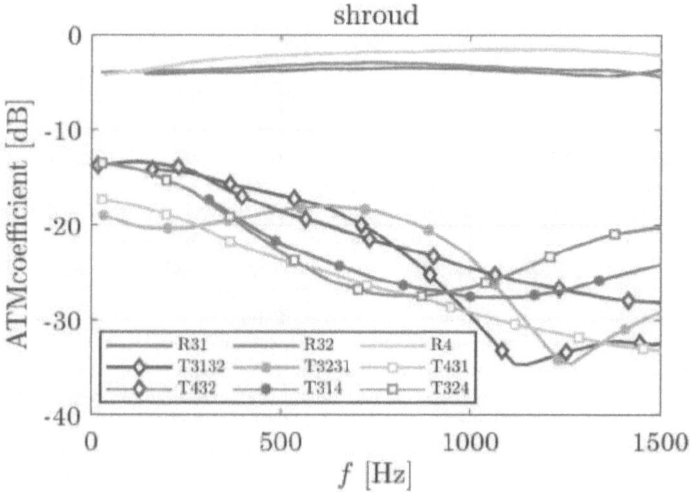

Figure 5.21: Components of the turbine acoustic transfer matrix, pulsating in shroud branch.

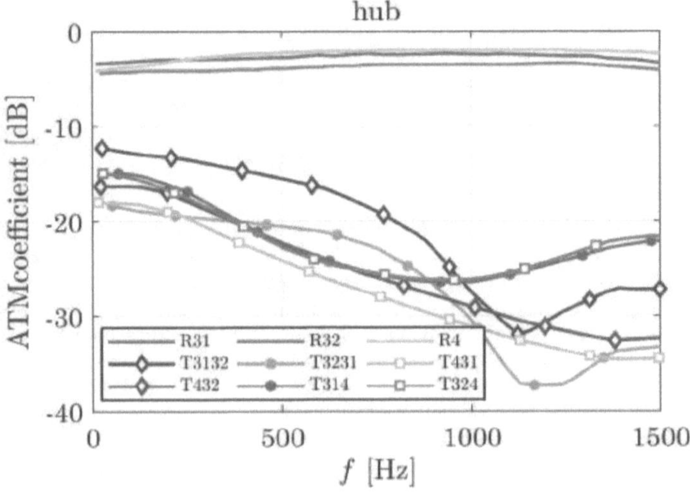

Figure 5.22: Components of the turbine acoustic transfer matrix, pulsating in hub branch.

In Figure 5.23 and Figure 5.24 the reflected, indirect reflected and transmitted pressure components are plotted for both experiments as well as the acoustic transfer matrix approximation for pulsating in shroud and hub respectively. For clarity's sake, the experimental case presented as an example is precisely the one selected in Figure 5.21 and Figure 5.22 for analysing the acoustic transfer matrix components. This way, it is straightforward to understand how the evolution of the acoustic transfer matrix components translates into the corresponding evolution of the reflected, indirect transmitted and transmitted pressure components. In general, it can be stated that the method is able to reproduce the acoustics in the lower range of frequencies, whereas, for higher than 500 Hz, the discrepancies between model and experiments start to increase significantly. Furthermore, although in general the results where similar for all the experimental points, a slightly increase of the error takes place for MFR's more separated of the symmetric MFR 0.5. This might indicate than the mass flow unbalance between the turbine inlet branches introduces some non-linearities that makes more inaccurate the acoustic transfer matrix approximation.

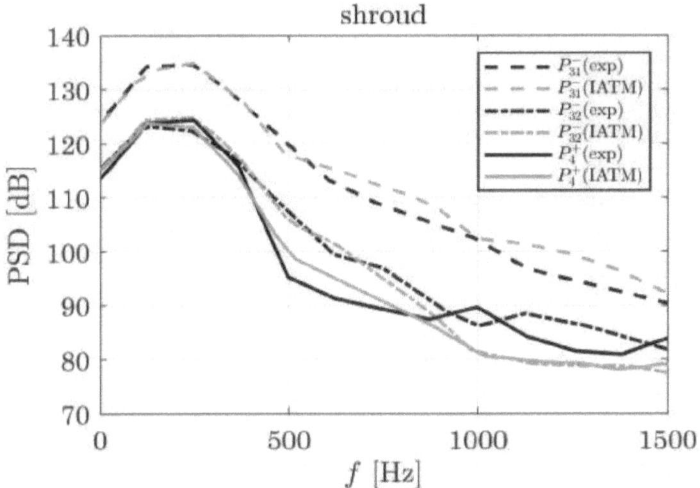

Figure 5.23: Experiments versus model for pulsating in shroud.

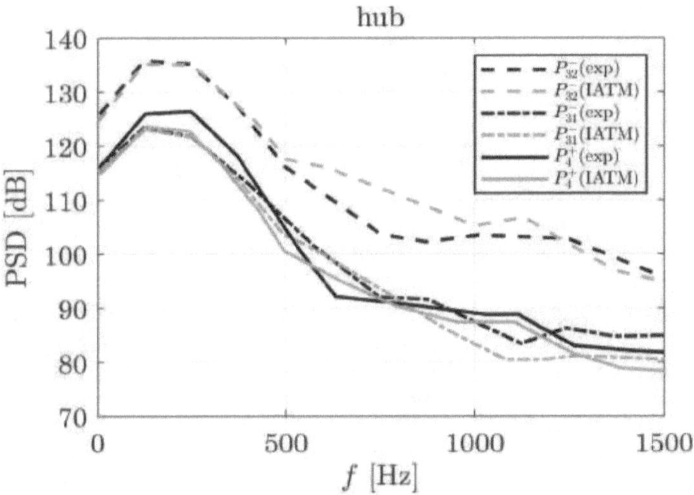

Figure 5.24: Experiments versus model for pulsating in hub.

Finally, for the next study, a synthetic pulse with higher amplitude, MFR 0.5 and 100 krpm, is used to explore the limitations of the application. First, the results that the method provide while computing the acoustic transfer matrix in MFR 0.5 are compared to the ones provided by the pure 1D model. Secondly, a comparison is made against the pure 1D model, but, in this case, interpolating a weighted averaged computation of the acoustic transfer matrices in MFR 0, MFR 0.5 and MFR 1, as it will be detailed in the following paragraphs.

In Figure 5.25, the synthetic pulse designed for the next studies is presented. As it can be observed, the pulse amplitude is approximately 1 bar and the pulses are originated in each of the inlet branches alternatively, which will certainly introduce strong oscillations around the MFR 0.5 level.

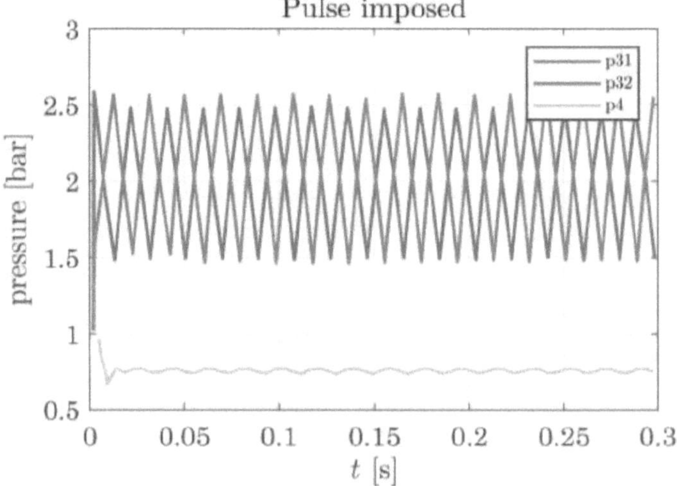

Figure 5.25: Synthetic pulse.

Following the procedure exposed in the previous results, i.e., in Figure 5.23 and Figure 5.24, the pressure decomposition results have been obtained using the acoustic transfer matrices computed for MFR 0.5, which coincides with the mean MFR of the synthetic pulse. The comparison against the pressure decompositions results provided by the pure 1D model are shown in Figure 5.26. At this point, it can be defended that, as it has been proved through the different validation campaigns along this book, the pure 1D model is able to reproduce the pulsating conditions in the turbine even for high amplitudes. Thus, if the acoustic transfer method presents a similar behaviour along the frequency domain, it would be a good indicator of the reliability of the approximation.

Indeed, taking into account that only the MFR 0.5 set of matrices has been used, results for the inlet branches are quite near the pure 1D ones until the 1000 Hz level. However, the pressure evolution in the outlet branch is not well reproduced even in the lower frequency range, which might imply that the method is particularly sensible to this high amplitude oscillation when trying to predict the turbine outlet conditions.

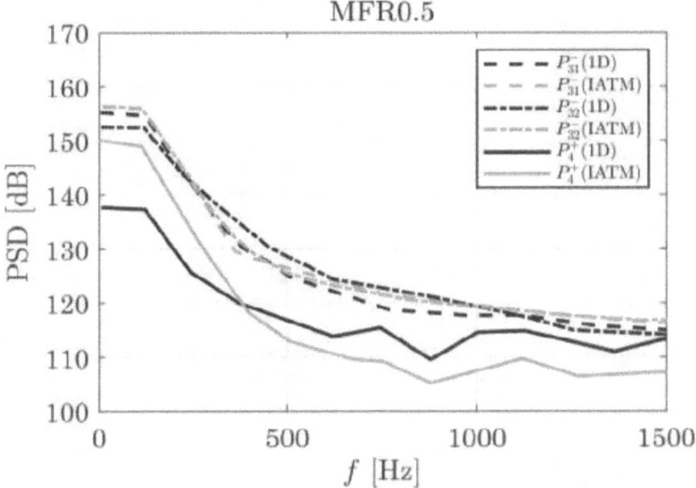

Figure 5.26: Comparison between pure 1D model and IATM computed in MFR 0.5.

Finally, for the next study, also the acoustic transfer matrices of MFR 0 and MFR 1 are computed in order to obtain a weighted averaged acoustic transfer matrix (ATM_{avg}). As it can be observed in Equation 5.13, it has been decided to overweight the ATM calculated in the MFR 0.5 due to the fact that this value is more frequently reached for the specific synthetic pulse used.

$$ATM_{\text{avg}} = 0.25 \cdot ATM_{\text{MFR}=0} + 0.25 \cdot ATM_{\text{MFR}=1} + 0.5 \cdot ATM_{\text{MFR}=0.5} \qquad (5.13)$$

In Figure 5.27, the comparison between the pure 1D model and the results using the ATM_{avg} are presented. From the graph it can be concluded that using an averaged acoustic transfer matrix is not an improvement, as results are very similar to the ones obtained using only the MFR 0.5.

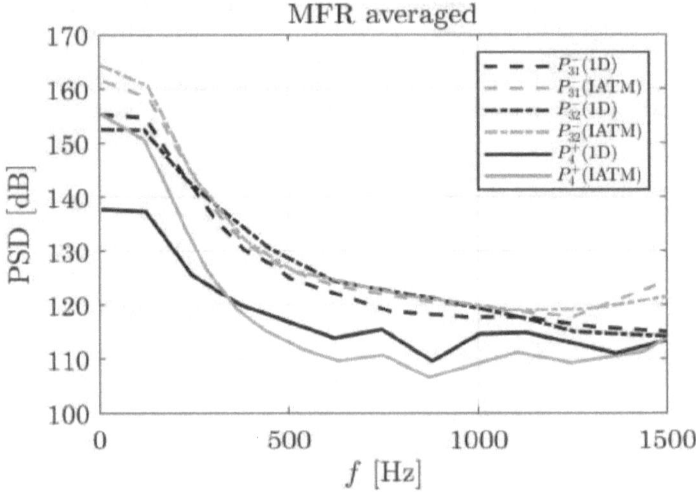

Figure 5.27: Comparison between pure 1D model and IATM computed for a weighted averaged of MFR 0, MFR 0.5 and MFR 1.

5.5   Conclusions

There is a clear time-scale separation between the very low frequency changes in the operating conditions of a radial turbine coupled to an engine (i.e., changes in engine load and rotational speed) and the instantaneous evolution of the pressure and temperature pulses due to the opening and closing of the exhaust valves. This way, the turbine behaviour can be approximated by that of an acoustic two-ports element superimposed to its average performance at each engine cycle. The acoustic two-ports element characteristics can be obtained from one-dimensional simulations and saved in a lookup table. These characteristics, in the form of an acoustic transfer matrix, can be interpolated each engine cycle in engine meanline models, with a very reduced computational cost. The generation of the lookup table, which is trivially parallelizable, can be performed in a preprocessing phase, and can be saved for later usage.

The accuracy of an acoustic matrix lookup table obtained with this procedure is directly related to the accuracy of the one-dimensional simulation it is based on. Good turbine map extrapolation capabilities are needed, as well as good acoustics-resolving one-dimensional computations. From the results obtained in this chapter, the number of points needed to get precise interpolations with the lookup table is fairly small, and should not be a big concern when applying this method.

The acoustic performance of a radial variable nozzle turbine seems to be mainly affected by its effective section. As the main parameter affecting the effective section is the position of its vanes, most of the changes in the acoustic performance are explained by it. The turbine expansion ratio is the second most important term affecting the effective section, whereas the reduced speed mainly affects the effective section at high blade speed ratios. In that off-design conditions, with little mass flow, the very low adiabatic efficiency of the turbine impacts the acoustic behaviour, producing a drift from that of a simple restriction between two ducts. This way, a reliable extrapolation of the adiabatic efficiency becomes mandatory when the sound and noise emissions of a radial turbine are to be computed at low engine loads and, thus, low expansion ratios.

It is worth mentioning that these kind of modelling only takes into account the passive acoustics of radial turbines. More work is needed if the active noise generation is needed, or if results are needed for frequencies outside of the plane wave range. Finally, specific experiments are needed to evaluate the sound and noise emissions and validate the model at extreme off-design conditions.

Finally, preliminary results of the method for the twin-entry case indicate that, although promising, additional difficulties appear and, thus, further studies should be made. More specifically, the turbine effective areas are strongly modified with the MFR, introducing non-linearities intrinsic to the interaction between the flow coming from the different inlet branches. Therefore, utilising a single MFR acoustic matrix is not enough and additional set of acoustic matrix should be computed. Also a method to interpolate in this additional dimension of MFR is to be designed and tested.

Chapter 5 References

[1]   J. Galindo, F. J. Arnau, L. M. García-Cuevas, and P. Soler. "Experimental validation of a quasi-two-dimensional radial turbine model". *International Journal of Engine Research* (2018). ISSN: 1468-0874. DOI: 10.1177/1468087418788502 (cit. on pp. xi, 8, 57, 62, 84, 90, 113, 131, 133, 139, 177).

[2]   J. Serrano, F. Arnau, L. M. García-Cuevas, P. Soler, L. Smith, R. Cheung, and B. Pla. "An Experimental Method to Test Twin and Double Entry Automotive Turbines in Realistic Engine Pulse Conditions". In: *WCX SAE World Congress Experience*. SAE Technical Paper 2019-01-0319. 2019. DOI: 10.4271/2019-01-0319 (cit. on pp. xi, 11, 100, 130).

[17] CMT – Motores T'ermicos, Universitat Politècnica de València. *OpenWAM*. 2016. URL: http://www.openwam.org/ (cit. on pp. 7, 11, 22, 98, 133, 178).

[25] J. R. Serrano, F. J. Arnau, V. Dolz, A. Tiseira, and C. Cervelló. "A model of turbocharger radial turbines appropriate to be used in zero- and one-dimensional

gas dynamics codes for internal combustion engines modelling". *Energy Conversion and Management* 49(12) (2008), pp. 3729 –3745. ISSN: 0196-8904. DOI: 10.1016/j.enconman.2008.06.031 (cit. on pp. 8, 30, 31, 96, 99, 100, 134).

[27] F. Payri, P. Olmeda, F. J. Arnau, A. Dombrovsky, and L. Smith. "External heat losses in small turbochargers: Model and experiments". *Energy* 71 (2014), pp. 534 –546. ISSN: 0360-5442. DOI: 10.1016/j.energy.2014.04.096 (cit. on pp. 8, 20, 27, 92, 131).

[28] J. R. Serrano, P. Olmeda, F. J. Arnau, A. Dombrovsky, and L. Smith. "Turbocharger heat transfer and mechanical losses influence in predicting engines performance by using one-dimensional simulation codes". *Energy* 86 (2015), pp. 204 –218. DOI: 10.1016/j.energy.2015.03.130 (cit. on pp. 8, 27, 92, 131).

[29] A. Gil, A. Tiseira, L. M. García-Cuevas, T. Rodríguez Usaquén, and G. Mijotte. "Fast three-dimensional heat transfer model for computing internal temperatures in the bearing housing of automotive turbochargers". *International Journal of Engine Research* (2018). DOI: 10.1177/1468087418804949 (cit. on pp. 8, 27, 92, 131).

[30] J. R. Serrano, F. J. Arnau, L. M. García-Cuevas, A. Dombrovsky, and H. Tartoussi. "Development and validation of a radial turbine efficiency and mass flow model at design and off-design conditions". *Energy Conversion and Management* 128 (2016), pp. 281 –293. ISSN: 0196-8904. DOI: 10.1016/j.enconman.2016.09.032 (cit. on pp. 8, 26–29, 96, 98, 131, 134, 135, 139, 157, 178).

[31] J. Galindo, R. Navarro, L. M. García-Cuevas, D. Tarí, H. Tartoussi, and S. Guilain. "A zonal approach for estimating pressure ratio at compressor extreme off-design conditions". *International Journal of Engine Research* 20(4) (2019), pp. 393–404. DOI: 10.1177/1468087418754899 (cit. on pp. 8, 92, 131).

[32] J. R. Serrano, P. Olmeda, A. Tiseira, L. M. García-Cuevas, and A. Lefebvre. "Theoretical and experimental study of mechanical losses in automotive turbochargers". *Energy* 55(0) (2013), pp. 888 –898. ISSN: 0360-5442. DOI: 10.1016/j.energy.2013.04.042 (cit. on pp. 8, 21, 92, 131).

[33] J. Galindo, H. Climent, A. Tiseira, and L. M. García-Cuevas. "Effect of the numerical scheme resolution on quasi-2D simulation of an automotive radial turbine under highly pulsating flow". *Journal of Computational and Applied Mathematics* 291 (Jan. 2016), pp. 112–126. DOI: 10.1016/j.cam.2015.02.025 (cit. on pp. 8, 131).

[35] A. W. Costall, R. M. McDavid, R. F. Martínez-Botas, and N. C. Baines. "Pulse performance modelling of a twin-entry turbocharger turbine under full unequal admission". In: *Proceedings of ASME Turbo Expo 2009*. 2009. ASME, 2009. DOI: 10.1115/1.4000566 (cit. on pp. 8, 20, 130).

[79] J. Galindo, P. Fajardo, R. Navarro, and L. M. García-Cuevas. "Characterization of a radial turbocharger turbine in pulsating flow by means of CFD and its application to engine modeling". *Applied Energy* 103(0) (2013), pp. 116 –127. ISSN: 0306-2619. DOI: 10.1016/j.apenergy.2012.09.013 (cit. on pp. 21, 56, 90, 130).

[86] K. G. Hohenberg, P. J. Newton, R. F. Martinez-Botas, M. Halamek, K. Maeda, and J. Bouilly. "Development and Experimental Validation of a Low Order Turbine Model Under Highly Pulsating Flow". In: *Turbo Expo: Power for Land, Sea, and Air, Volume 2C: Turbomachinery*. ASME. June 2017, V02CT44A023. DOI: 10.1115/GT2017-63983 (cit. on pp. 21, 57, 90, 130).

[87] J. R. Serrano, A. Tiseira, L. M. García-Cuevas, L. Inhestern, and H. Tartoussi. "Radial turbine performance measurement under extreme off-design conditions". *Energy* 125 (2017), pp. 72–84. DOI: 10.1016/j.energy.2017.02.118 (cit. on pp. 21, 57, 90, 135).

[88] A. J. Torregrosa, A. Broatch, R. Navarro, and J. García-Tíscar. "Acoustic characterization of automotive turbocompressors". *International Journal of Engine Research* 16(1) (2015), pp. 31–37. DOI: 10.1177/1468087414562866 (cit. on pp. 21, 57, 90, 130).

[90] J. Galindo, A. Tiseira, R. Navarro, D. Tarí, and C. Meano. "Effect of the inlet geometry on performance, surge margin and noise emission of an automotive turbocharger compressor". *Applied Thermal Engineering* 110 (2017), pp. 875–882. DOI: 10.1016/j.applthermaleng.2016.08.099 (cit. on pp. 21, 130).

[91] J. Galindo, A. Tiseira, P. Fajardo, and L. M. García-Cuevas. "Development and validation of a radial variable geometry turbine model for transient pulsating flow applications". *Energy Conversion and Management* 85 (2014), pp. 190 –203. ISSN: 0196-8904. DOI: 10.1016/j.enconman.2014.05.072 (cit. on pp. 21, 25, 131).

[93] Z. Ding, W. Zhuge, Y. Zhang, H. Chen, R. Martinez-Botas, and M. Yang. "A one-dimensional unsteady performance model for turbocharger turbines". *Energy* (2017). ISSN: 0360-5442. DOI: 10.1016/j.energy.2017.04.154 (cit. on pp. 21, 130, 131).

[94] J. Galindo, J. R. Serrano, F. J. Arnau, and P. Piqueras. "Description of a Semi-Independent Time Discretization Methodology for a One-Dimensional Gas Dynamics Model". *Journal of Engineering for Gas Turbines and Power* 131(3) (2009), p. 034504. ISSN: 07424795. DOI: 10.1115/1.2983015 (cit. on pp. 23, 133).

[98] J. R. Serrano, F. J. Arnau, L. M. García-Cuevas, and L. B. Inhestern. "An innovative losses model for efficiency map fitting of vaneless and variable vaned radial turbines extrapolating towards extreme off-design conditions". *Energy* 180 (2019), 626 –639. ISSN: 0360-5442. DOI: 10.1016/j.energy.2019.05.062 (cit. on pp. 26, 157).

[101] G Pinero, L Vergara, J. M. Desantes, and A Broatch. "Estimation of velocity fluctuation in internal combustion engine exhaust systems through beamforming techniques". *Measurement Science and Technology* 11(11) (2000), p. 1585. DOI: 10.1088/0957-0233/11/11/307 (cit. on pp. 36, 39, 64, 133, 177).

[102] P. Welch. "The Use of Fast Fourier Transform for the Estimation of Power Spectra: A Method Based on Time Averaging Over Short, Modified Periodograms". *IEEE Transactions on Audio Electroacoustics* 15 (2 1967), pp. 70–73. ISSN: 0018-9278. DOI: 10.1109/TAU.1967.1161901 (cit. on pp. 39, 106, 136).

[120] D. Winterbone and R. Pearson. "Turbocharger turbine performance under unsteady flow-a review of experimental results and proposed models". *Inst Mech Eng Semin, paper C55403198* (1998) (cit. on pp. 90, 130).

[121] F. Piscaglia, A. Onorati, S. Marelli, and M. Capobianco. "A detailed one-dimensional model to predict the unsteady behavior of turbocharger turbines for internal combustion engine applications". *International Journal of Engine Research* 20(3) (2019), pp. 327–349. DOI: 10.1177/1468087417752525 (cit. on pp. 90, 130).

[134] H. Rämmal and M. Åbom. "Acoustics of Turbochargers". In: *SAE 2007 Noise and Vibration Conference and Exhibition*. SAE International, 2007. DOI: 10.4271/2007-01-2205 (cit. on p. 130).

[135] A. Broatch, J. Galindo, R. Navarro, and J. García-Tíscar. "Methodology for experimental validation of a CFD model for predicting noise generation in centrifugal compressors". *International Journal of Heat and Fluid Flow* 50 (2014), 134–144. DOI: 10.1016/j.ijheatfluidflow.2014.06.006 (cit. on p. 130).

[136] A. Broatch, X. Margot, J. García-Tíscar, and F. Roig. "Impact of simple surge-enhancing inlet geometries on the acoustic behavior of a turbocompressor". *International Journal of Engine Research* (2018). DOI: 10.1177/1468087418784125 (cit. on p. 130).

[137] S. Sharma, A. Broatch, J. García-Tíscar, J. M. Allport, and A. K. Nickson. "Acoustic characteristics of a ported shroud turbocompressor operating at design conditions". *International Journal of Engine Research* (2018). DOI: 10.1177/1468087418814635 (cit. on p. 130).

[138] K. Peat, A. Torregrosa, A. Broatch, and T. Fernández. "An investigation into the passive acoustic effect of the turbine in an automotive turbocharger". *Journal of Sound and Vibration* 295 (1-2 Aug. 2006), pp. 60–75. DOI: 10.1016/j.jsv.2005.11.033 (cit. on p. 130).

[139] A. Torregrosa, J. Galindo, J. R. Serrano, and A. Tiseira. "A Procedure for the Unsteady Characterization of Turbochargers in Reciprocating Internal Combustion Engines". In: *Fluid Machinery and Fluid Mechanics*. Ed. by J. Xu, Y. Wu, Y. Zhang, and J. Zhang. Berlin, Heidelberg: Springer Berlin Heidelberg, 2009, pp. 72–79. DOI: 10.1007/978-3-540-89749-1_10 (cit. on p. 130).

[140] R. Kabral, Y. A. El Nemr, C. Ludwig, R. Mirlach, P. Koutsovasilis, A. Masrane, and M. Åbom. "Experimental acoustic characterization of automotive twin-scroll turbine". In: *12th European Conference on Turbomachinery Fluid dynamics & Thermodynamics*. European Turbomachinery Society. DOI: `10.29008/ETC2017-363` (cit. on p. 130).

[141] A. Broatch, J. Galindo, R. Navarro, J. García-Tíscar, A. Daglish, and R. K. Sharma. "Simulations and measurements of automotive turbocharger compressor whoosh noise". *Engineering Applications of Computational Fluid Mechanics* 9(1) (2015). DOI: `10.1080/19942060.2015.1004788` (cit. on p. 130).

[142] A. Broatch, J. Galindo, R. Navarro, and J. García-Tíscar. "Numerical and experimental analysis of automotive turbocharger compressor aeroacoustics at different operating conditions". *International Journal of Heat and Fluid Flow* 61 (2016), pp. 245–255. DOI: `10.1016/j.ijheatfluidflow.2016.04.003` (cit. on p. 130).

[143] A. Marsan and S. Moreau. "Analysis of the flow structure in a radial turbine". In: *11th European Conference on Turbomachinery Fluid Dynamics and Thermodynamics, ETC 2015*. Madrid, Spain, Mar. 2015 (cit. on p. 130).

[144] F. J. Wallace and J. Adgey. "Paper 1: Theoretical Assessment of the Non-Steady Flow Performance of Inward Radial Flow Turbines". *Proceedings of the Institution of Mechanical Engineers, Conference Proceedings* 182(8) (1967), pp. 22–36. DOI: `10.1243/PIME_CONF_1967_182_211_02` (cit. on p. 130).

[145] M. S. Chiong, S. Rajoo, A. Romagnoli, and R. Martínez-Botas. "Unsteady performance prediction of a single entry mixed flow turbine using 1-D gas dynamic code extended with meanline model". In: *Proceedings of the ASME Turbo Expo*. Vol. 5. ASME. 2012, pp. 781–795. DOI: `10.1115/GT2012-69176` (cit. on p. 131).

[146] C. Avola, C. Copeland, A. Romagnoli, R. Burke, and P. Dimitriou. "Attempt to correlate simulations and measurements of turbine performance under pulsating flows for automotive turbochargers". *Proceedings of the Institution of Mechanical Engineers, Part D: Journal of Automobile Engineering* 233(2) (2019), pp. 174–187. DOI: `10.1177/0954407017739123` (cit. on p. 131).

[147] R. Veloso, Y. Elnemr, F. M. Reich, and S. Allam. "Simulation of Sound Transmission through Automotive Turbochargers". In: *7th International Styrian Noise, Vibration & Harshness Congress: The European Automotive Noise Conference*. SAE International, June 2012. DOI: `10.4271/2012-01-1560` (cit. on p. 131).

[148] J. Galindo, A. Tiseira, R. Navarro, D. Tarí, H. Tartoussi, and S. Guilain. "Compressor Efficiency Extrapolation for 0D-1D Engine Simulations". In: *SAE 2016 World Congress and Exhibition*. SAE Technical Paper 2016-01-0554. SAE International, 2016. DOI: `10.4271/2016-01-0554` (cit. on p. 131).

[149] Z. Zhu, S. Midlam-Mohler, and M. Canova. "Development of physics-based three-way catalytic converter model for real-time distributed temperature prediction using proper orthogonal decomposition and collocation". *International Journal of Engine Research* (2019). DOI: `10.1177/1468087419876127` (cit. on p. 150).

[150] D. Gordon, C. Wouters, M. Wick, F. Xia, B. Lehrheuer, J. Andert, C. R. Koch, and S. Pischinger. "Development and experimental validation of a real-time capable field programmable gate array–based gas exchange model for negative valve overlap". *International Journal of Engine Research* (2019). DOI: 10.1177/1468087418788491 (cit. on p. 150).

Chapter 6

Concluding remarks

Contents

6.1 Introduction

In the present work, a twin-entry turbine model for non-linear pulsating simulations is presented, along with an experimental validation of the quasi-2D turbine model first presented in [53], both able to be integrated in the existing one-dimensional engine simulation codes. Furthermore, the method also here developed for fast 1D turbine sound characterisation allows the models to be deployed even for some applications where very low computational cost is needed. In order to validate the models it has been necessary to develop a new experimental procedure in which it is feasible to reach a wide range of turbine operating conditions regardless of the specific two-scroll turbine type.

The current chapter reviews the outcome of this book. The main findings and contributions are presented in section 6.2 for both experiments and modelling. Then, section 6.3 enumerates the limitations of the model approach and experimental procedures that have been followed. Finally, section 6.4 includes the potential improvements suggested and possible lines of work that may be explored for further research.

6.2 Main contributions

This book has contributed to the understanding and performance prediction capabilities of the radial turbine machine for automotive applications, mainly by studying the always challenging non-linear pulsating flow. The new modelling tools developed are specifically designed to account for the non-linear effects and complex phenomena that are inescapable when analysing realistic operating conditions, without neglecting that keeping low the computational cost is still crucial for some applications. Also, a new procedure to experimentally test any two scroll turbine under any realistic operating conditions is presented, with the intention of not only validating the models here developed, but also contributing to the effort of many other researches that confront the challenge of finding flexible and easily implementable methods to test this complex machines under a wide variety of conditions.

But not only this book implied the creation of new tools, but also several interesting findings have emerged from the validation campaign of the models. These findings that come from the results of the studies carried out using the model give a full illustration of how the tools here developed might modestly contribute to the design of more efficient turbocharger or aftertreatment systems as well as even lead to a better turbocharger operating range selection and matching. The tools, methods and procedures that compose this book have been elaborated taking into fully consideration the practical applicability of them, and, thus, the feasibility to integrate them in real industrial projects. On the one hand, the implementation of the model in the code OpenWAM is compatible to be used in co-simulation with some of the software used in the industry as de-facto standards, such as GT-POWER and Ricardo WAVE. Furthermore, the acoustic transfer matrix interpolation method (IATM) is specifically thought to be used in the increasingly demanded real-time industry applications. The highlighted contributions of the current document are presented next, being divided into three main categories: experiments, modelling work and industrial applications.

6.2.1 Experimental procedure

The experimental work of this book has been carried out using the CMT-Motores Térmi-cos gas stand laboratory, a versatile facility for testing any type of turbocharger isolated from the IC engine in a wide range of operating conditions. Using previous experience [1] and applying contrasted techniques [101], the gas stand has been successfully configured to be able to be utilised in experimental campaigns concerning any type of two-scroll turbocharger under both hot and cold flow conditions and steady and pulsating flow conditions. In addition, when designing the installation proposed in this work, the fo-cus was placed in the capability of obtaining big amplitude engine-like pulses. In this regard, the amplitude reached the 150-200 mbar range in the experiments presented in chapter 3, which, although not as high as the amplitude reached in a pulse generated in a real engine, it is high enough to enter in the non-linearity domain, as it has been analytically demonstrated in chapter 3.

Obtaining reliable and accurate experimental measurements when dealing with un-steady flow in a very high speed rotating machine is a huge challenge, but the acquisition of realistic turbocharger high quality data is increasingly important in the actual context of ICE research and development, where, apart from being necessary for validating the models, it is also crucial for a better understanding of a complex machine that plays a very important role in the ever-rising importance of the final global efficiency of the engine. When deepening into the data from the experiments, some interesting findings related to the behaviour of the flow near the turbine under non-linear pulses have emer-ged. As an illustrative example, experiments reveal that, although two-scroll turbines enable a superior control of the engine pulses, significant or higher than expected oscilla-tions are still transmitted from the pulsating (active) inlet branch to the other (passive) inlet branch which, in addition to the energy travelling due to the reflection component, might significantly affect the pumping losses and total fuel consumption.

6.2.2 Modelling

The modelling contributions exposed in this book fall into three lines of work. Firstly, a quasi-2D turbine model continuing the work from [53] has been experimentally validated as seen in chapter 2. Secondly, a new twin-entry radial turbine model able to be used in non-linear pulsating simulations is developed in chapter 4 that comple-ments the two-scroll turbine extrapolation model presented in [34]. Finally, the fast one-dimensional model based on acoustic matrices from chapter 5 provides a method that is mainly conceived to provide the reliability and accuracy of a one-dimensional turbine model like the ones presented in this book, but adding real time capabilities.

- **Quasi-2D radial turbine model**. This model was first presented in [53], where it proved to be a robust method to improve the performance of the equivalent actual one-dimensional duct approaches particularly at high frequencies, without a significant penalty in the computational cost. The contributions of this book cover the integration and validation of the model in the one-dimensional code OpenWAM [17]. Thus, providing it with the necessary features that allow a compa-tibility with the other submodels that where previously implemented in the code, such as turbocharger heat transfer or mechanical losses models as well as with the flow boundary conditions based on Godunov's method [95]. In addition, in this work the quasi-2D model is first time validated with the new turbine extrapola-tion model [30]. With the validation from chapter 2 against instantaneous pressure data, it becomes clear how the classic volute model produces good results at up to 800-1000 Hz, while the quasi-2D model improved these predictions in the higher range above 1000 Hz and even close to the level of 2000 Hz.

- **Twin-entry radial turbine model**. When developing this model, the major effort was placed in solving the difficulty that accounting for non-linear effects implies when trying to predict the turbine performance under highly pulsating conditions. The designed structure composed of three equivalent nozzles and an intermediate volume is able to reproduce with high accuracy the main sources of linear and non-linear acoustics effects of the turbine. Using a small delay for the computation of the MFR turned out to be an effective solution to the instabilities of the model when shifting from steady to unsteady simulations while keeping its computation explicit. Regarding the results of the validation, it can be stated that the model successfully captures the non-linear phenomena for the medium and medium-high frequency spectra. Thus, this method could provide a useful tool when trying to predict the effect of the interferences transmitted during the exhaust process that, according to the experiments performed in chapter 3, might be higher enough to affect the global engine performance. In addition, it is worth noting that the model is implemented with the necessary modularity and flexibility to enable co-simulation campaigns. In this regard, the software was implemented as a plugin for both GT-POWER and Ricardo WAVE, leading to results comparable to that exposed in chapter 3.

- **Acoustic transfer matrices method**. Estimating correctly the turbine acoustics can be valuable during the engine design stage, in fact, it can lead to a more optimised design of the silencer and aftertreatment, as well as to a better prediction of the scavenging effects. Results developed throughout this book have demonstrated that the one-dimensional approaches provides good prediction capabilities without the computational cost that a more detailed model , such as CFD models, implies. To obtain the sound and noise emissions of radial turbocharger turbines with even lower computational costs is a challenge. However, the increasing interest in applications requiring real time capabilities indicates that it might be worth exploring solutions to asses this demand. In chapter 5, a method is proposed in which the acoustic response of single entry radial turbines can be characterised by means of acoustic transfer matrices that change with the operating conditions. The first applications of this model show that a simple interpolation of lookup tables of acoustic transfer matrices combines one-dimensional prediction with real time capabilities when being used in mean-value simulations campaigns. In the case of twin-scroll turbines, the investigations here exposed also indicate that adapting the method for such applications is not straight-forward and additional data might be fed into the model due to the complex interactions that the mixed flow implies.

6.2.3   Applications and validation process

After the design and implementation phases of the model and the set up of the boundary conditions, the full potential of the tools is still unknown. The development process of the model would only be complete once an extensive validation is carried out, whether to show its performance and accuracy or to provide examples where it can be applied. To this end, the matrices of experimental points for both single entry and twin-entry experiments have been selected covering a wide range of turbine operating conditions, testing several turbocharger rotational speeds and VGT positions or mass flow ratios (MFR). Concerning the frequency and amplitude of the pulses applied, the former are selected considering various engine equivalent speeds, whereas the latter are lower enough to maintain controlled conditions during testing, but higher enough to assure realistic non-linearity conditions. On the other hand, once the models have been previously validated, in chapter 5, the focus is placed on demonstrating the viability of characterising the turbine acoustics by interpolating transfer matrices, and the computational cost that this procedure would imply if intended to be used for real time applications.

The results of the validations of the models indicate that it is possible to capture the non-linear effects until the medium and even part of the high frequency domain, reproducing the acoustic response sometimes even in the range of 1500-2000 Hz. From the analysis of the instantaneous tests, it is clear that during the exhaust process, performance variables such as mass flow, efficiency, MFR, etc. are highly modified due to the oscillations transmitted. Predicting and simulating this instantaneous variations might enable wiser decisions in terms of design of exhaust line components or turbocharger range of operation selection.

Nevertheless, the objective of the work carried out in this book is not only to contribute to the understanding of the unsteady performance of radial turbines by means of developing new models, but also to ensure the applicability of them. For this purpose, the models have been implemented with the appropriate flexibility and modularity, with the aim of providing a easy applicable piece of software able to be used as an embedded tool in commercial software. For its part, the fast simulation method from chapter 5 is intended to satisfy the demand for simulation tools with real time capabilities, although limitations have been found when utilising this method in two-scroll turbine applications.

6.3 Limitations

Even though the main objectives stated in chapter 1 have been accomplished, the models developed present some limitations that will be next discussed. Also the pulsating experiments procedure for two-scroll turbines has its own restrictions that might be overcome in future revisions of the method.

First, although the quasi-2D introduces more detail in the representation of the volute geometry, the improvement in the prediction capabilities seems to be restricted to a specific range of the middle-high frequency domain. With the validation carried out, it is not prudent to affirm that the quasi-2D model improves the prediction of the instantaneous response of turbine operating parameters such as efficiency and, thus, further studies should be made. As a consequence, for most applications it might not be worthwhile the extra computational cost that the model implies. However, this hassle can be mitigated if the model is complemented with the acoustic transfer interpolation method presented in chapter 5 so the computational cost would fall into the preprocessing step. On the other hand, although it can be stated that the solutions adopted for simulating the twin-scroll turbine, such as the intermediate volume have a physical explanation, when applied to other two-scroll geometries, some instabilities might appear and so, modifications and additional calibrations might be necessary.

Regarding the experiments, it is important to mention the limitations in relation to the amplitude and specific shape of the pulses that can be performed using the facility and methodology presented in this book. The system of orifice plate and rotating valve provides a stable and secure solution for performing pulses and it allows to reproduce the frequency of an equivalent engine of any configuration of cylinders. However, the specific pulsating gas dynamic oscillation entering the engine air system is limited to an specific wave shape whose nature is generated by its pass through the orifice. A more modular design philosophy as the Transient Air System Rig [109] would provide more flexibility in terms of analysing the response to different wave shapes. Due to limitations in the facility and the instrumentation used, the maximum amplitude of the pulse was also limited and, although it was higher enough to introduce non-linearity, it introduces some limitations in order to fully recreate the transient and pulsating gas dynamics from a real internal combustion engine.

Finally, the acoustic transfer matrices method has proven to be a very smart tool to perform real time mean value simulations but with the advantage of fully one-dimensional accuracy. In fact, the results using the interpolation method for the single entry turbine are extremely close to the ones provided by the 1D model used to create the interpolation matrices, in addition, the computational cost has successfully declined to be much lower than the level required for real time applications. However, the method presents some evident restrictions when applied to two-scroll turbines. The MFR strongly modifies the equivalent effective area by introducing non-linearities generated by the interaction of the flow coming from the different entries that are added to the non-linearities intrinsic to the pulses. In sum, this phenomena difficulties the interpolation procedure and introduces the necessity of feeding the model with extra data, as well as to probably evolve the interpolation method to a slightly more complex procedure.

6.4 Suggestions for future studies

Each of the scientific contributions and lines of work in which the respondent has participated during the elaboration of this book offers the possibility of further studies and improvements.

The experimental procedure exposed in chapter 3 has been devised from the beginning as a flexible method able to be used with any specific type of two-scroll turbocharger. In the moment of the publication of this book, the respondent's research team has performed experimental tests in steady state conditions for a double-entry turbine in the same gas stand facility as the one described in this work and also pulsating tests will be performed in the near future. These tests will allow to validate the experimental procedures that are described in this work with a different turbine configuration, and also offers the opportunity to improve some of the technical limitations that appeared during the twin turbine tests. More specifically, it will probably be feasible to increase the amplitude of the pulse by using shorter turbine pipes in a more optimised installation, as well as a with a better selection of the by-pass valve in order to avoid the leaks.

The quasi-bidimensional turbine volute model could be adapted to a compressor model. Each element, compressor volute and diffuser can be discretised into a row of cells which probably will enable more accurate simulations of the flow dynamics and a better prediction of the pulsating performance in the high frequencies. Furthermore, the flow reversal phenomena present during surge episodes might affect only part of the diffuser and, thus, would probably be easy to compute with this method. Future studies in this topic could also include a quasi-bidimensional version of the twin-entry and double-entry turbines, which is especially straight-forward in the twin-entry case, where each volute can be modelled as a row of cells connected to the stator nozzles.

From the experiments and simulations of the twin-entry turbine, there are several interesting trends and relations that have been observed. Among other variables, the geometry of the entry and the MFR seems to affect the transmitted and reflected losses, however, due to non-linearities introduced by the interaction of the flow coming from the different entries, it is difficult to identify the global tendencies. In order to better comprehend how these variables affect the turbine acoustics, Computational Fluid Dynamics (CFD) simulations should be performed, using different geometries of turbine entries and a set of unequal turbine conditions.

The procedure to adapt the acoustic transfer matrices method to the more complex two-scroll turbine has been already formulated in chapter 5. The equivalent equations for the transfer matrices have been proposed and also some results have been obtained. From these preliminary results a big part of the difficulties intrinsic to the application of the method have already arisen coming from the interaction of the two flow streams when dealing with the unequal or partial admissions. To confront this issue, one possible approach is to generate extra data of acoustic transfer matrices for different Mass Flow Ratios (MFR) as using only the MFR 0.5 matrix will give inaccurate results. After the generation of the data matrix, a weighted averaged interpolation procedure could be a good approach for approximating the acoustic transfer matrix of the MFR intermediate positions.

Chapter 6 References

[1]   J. Galindo, F. J. Arnau, L. M. García-Cuevas, and P. Soler. "Experimental validation of a quasi-two-dimensional radial turbine model". *International Journal of Engine Research* (2018). ISSN: 1468-0874. DOI: 10.1177/1468087418788502 (cit. on pp. xi, 8, 57, 62, 84, 90, 113, 131, 133, 139, 177).

[17]   CMT – Motores T'ermicos, Universitat Politècnica de València. *OpenWAM*. 2016. URL: http://www.openwam.org/ (cit. on pp. 7, 11, 22, 98, 133, 178).

[30]   J. R. Serrano, F. J. Arnau, L. M. García-Cuevas, A. Dombrovsky, and H. Tartoussi. "Development and validation of a radial turbine efficiency and mass flow model at design and off-design conditions". *Energy Conversion and Management* 128 (2016), pp. 281 –293. ISSN: 0196-8904. DOI: 10.1016/j.enconman.2016.09.032 (cit. on pp. 8, 26–29, 96, 98, 131, 134, 135, 139, 157, 178).

[34]   J. Serrano, F. Arnau, L. Garcéa-Cuevas, and V. Samala. "Development of flow oriented model for extrapolation and interpolation off-design performance of twin-entry and dual volute radial-inflow turbines working under different flow admission conditions". *Energy* (). Under review (cit. on pp. 8, 92, 99, 100, 177).

[53]   L. M. G.-C. González. "Experiments and Modelling of Automotive Turbochargers under Unsteady Conditions". PhD book. Universitat Politècnica de València, 2014 (cit. on pp. 9–11, 176–178).

[95]   S. K. Godunov. "A Difference Scheme for Numerical Solution of Discontinuous Solution of Hydrodynamic Equations". *Matematicheskii Sbornik* 47 (1959), pp. 271–306 (cit. on pp. 23, 178).

[101] G Pinero, L Vergara, J. M. Desantes, and A Broatch. "Estimation of velocity fluctuation in internal combustion engine exhaust systems through beamforming techniques". *Measurement Science and Technology* 11(11) (2000), p. 1585. DOI: 10.1088/0957-0233/11/11/307 (cit. on pp. 36, 39, 64, 133, 177).

[109] A. Costall, V. Cheong, H. Flora, A. Munasinghe, R. Ivanov, R. W. Kruiswyk, and J. R. McDonald. "Development of a Novel Transient-Pulsating Flow Rig for Engine Air System Research using GT-SUITE". In: *European GT Conference.* 2018 (cit. on pp. 57, 180).

Global references

[1] J. Galindo, F. J. Arnau, L. M. García-Cuevas, and P. Soler. "Experimental validation of a quasi-two-dimensional radial turbine model". *International Journal of Engine Research* (2018). ISSN: 1468-0874. DOI: 10.1177/1468087418788502 (cit. on pp. xi, 8, 57, 62, 84, 90, 113, 131, 133, 139, 177).

[2] J. Serrano, F. Arnau, L. M. García-Cuevas, P. Soler, L. Smith, R. Cheung, and B. Pla. "An Experimental Method to Test Twin and Double Entry Automotive Turbines in Realistic Engine Pulse Conditions". In: *WCX SAE World Congress Experience*. SAE Technical Paper 2019-01-0319. 2019. DOI: 10.4271/2019-01-0319 (cit. on pp. xi, 11, 100, 130).

[3] J. R. Serrano, F. J. Arnau, L. M. García-Cuevas, P. Soler, and R. Cheung. "Experimental validation of a one-dimensional twin-entry radial turbine model under non-linear pulse conditions". *International Journal of Engine Research* (2019), p. 146808741986915. ISSN: 1468-0874. DOI: 10.1177/1468087419869157 (cit. on pp. xi, 8).

[4] A. Torregrosa, L. M. García-Cuevas, L. B. Inhestern, and P. Soler. "Radial turbine sound and noise characterisation with acoustic transfer matrices by means of fast one-dimensional models". *International Journal of Engine Research* (2019), p. 1468087419889429. DOI: 10.1177/1468087419889429. eprint: https://doi.org/10.1177/1468087419889429 (cit. on pp. xi, 8).

[5] J. S. Gaffney and N. A. Marley. "The impacts of combustion emissions on air quality and climate - From coal to biofuels and beyond". *Atmospheric Environment* 43(1) (2009). Atmospheric Environment - Fifty Years of Endeavour, pp. 23–36. ISSN: 1352-2310. DOI: 10.1016/j.atmosenv.2008.09.016 (cit. on p. 3).

[6] J. Serrano. "Imagining the Future of the Internal Combustion Engine for Ground Transport in the Current Context". *Applied Sciences* 7(10) (2017), p. 1001. ISSN: 2076-3417. DOI: 10.3390/app7101001 (cit. on pp. 3, 56).

[7] P. IEA. *Global EV Outlook 2019*. 2019. URL: www.iea.org/publications/reports/globalevoutlook2019/ (cit. on p. 3).

[8]   European Parliament and Council of the European Union. "Regulation (EU) No 510/2011 of the European Parliament and of the Council of 11 May 2011 setting emission performance standards for new light commercial vehicles as part of the Union's integrated approach to reduce CO 2 emissions from light-duty vehicles (Text with EEA relevance)". In: *Official Journal of the European Union 54*. 2011. DOI: 10.3000/17252555.L_2011.145.eng (cit. on p. 3).

[9]   C. EEA. *Monitoring of CO2 emissions from passenger cars – Regulation (EC) No 443/2009*. 2019. URL: https://www.eea.europa.eu/data-and-maps/data/co2-cars-emission-16 (cit. on p. 3).

[10]   European Parliament and Council of the European Union. "Regulation (EC) No 715/2007 of the European Parliament and of the Council of 20 June 2007 on type approval of motor vehicles with respect to emissions from light passenger and commercial vehicles (Euro 5 and Euro 6) and on access to vehicle repair and maintenance information (Text with EEA relevance)". In: *Official Journal of the European Union 50*. 2007. DOI: 10.3000/17252555.L_2011.145.eng (cit. on p. 3).

[11]   European Parliament and Council of the European Union. "Regulation (EC) No 595/2009 of the European Parliament and of the Council of 18 June 2009 on type-approval of motor vehicles and engines with respect to emissions from heavy duty vehicles (Euro VI) and on access to vehicle repair and maintenance information and amending Regulation (EC) No 715/2007 and Directive 2007/46/EC and repealing Directives 80/1269/EEC, 2005/55/EC and 2005/78/EC (Text with EEA relevance)". In: *Official Journal of the European Union 52*. 2009. DOI: 10.3000/17252555.L_2011.145.eng (cit. on p. 3).

[12]   J. R. Serrano, H. Climent, P. Piqueras, and E. Angiolini. "Filtration modelling in wall-flow particulate filters of low soot penetration thickness". *Energy* 112 (2016), pp. 883–898. ISSN: 0360-5442. DOI: 10.1016/j.energy.2016.06.121 (cit. on p. 4).

[13]   P. Michel, A. Charlet, G. Colin, Y. Chamaillard, G. Bloch, and C. Nouillant. "Optimizing fuel consumption and pollutant emissions of gasoline-HEV with catalytic converter". *Control Engineering Practice* 61 (2017), pp. 198–205. ISSN: 0967-0661. DOI: 10.1016/j.conengprac.2015.12.010 (cit. on p. 5).

[14]   F. Millo, F. Mallamo, E. Pautasso, and G. Ganio Mego. "The Potential of Electric Exhaust Gas Turbocharging for HD Diesel Engines". In: *SAE 2006 World Congress & Exhibition*. SAE International, 2006. DOI: 10.4271/2006-01-0437 (cit. on p. 5).

[15]   D. Evans and A. Ward. "Minimizing Turbocharger Whoosh Noise for Diesel Powertrains". *SAE Technical Paper* 2005-01-2485 (2005). DOI: 10.4271/2005-01-2485 (cit. on p. 5).

[16]   B. WLTP facts. *What is WLTP and how does it work*. 2019. URL: https://wltpfacts.eu/what-is-wltp-how-will-it-work/ (cit. on p. 5).

[17] CMT – Motores T'ermicos, Universitat Politècnica de València. *OpenWAM*. 2016. URL: http://www.openwam.org/ (cit. on pp. 7, 11, 22, 98, 133, 178).

[18] J. M. Corberán. *Contribución al modelado del proceso de renovación de la carga en motores de combustión interna alternativos*. PHD book. 1984 (cit. on pp. 7, 8).

[19] R. Benson. "The thermodynamics and gas dynamics of internal-combustion engines". In: *Proceedings of the IEEE*. Vol. 66. IEEE, 1978, pp. 51–83. DOI: 10.1109/PROC.1978.10837 (cit. on p. 7).

[20] B van Leer. "Towards the ultimate conservative difference scheme, V. A second order sequel to Godunov's method". *Journal of Computational Physics* 32 (1979), pp. 101–136. DOI: 10.1016/0021-9991(79)90145-1 (cit. on p. 8).

[21] E. Toro, M. Spruce, and W. Speares. "Restoration of the contact surface in the HLL-Riemann solver". English. *Shock Waves* 4(1) (1994), pp. 25–34. ISSN: 0938-1287. DOI: 10.1007/BF01414629 (cit. on pp. 8, 23).

[22] F. Payri, J. Desantes, and J. Corberán. "A quasi-steady model on gas exchange process, some results". In: *Motor Sympo 88*. 1988 (cit. on p. 8).

[23] F. Payri, J. Benajes, and M. Reyes. "Modelling of supercharger turbines in internal-combustion engines". *International Journal of Mechanical Sciences* 38(8) (1996), pp. 853 –869. ISSN: 0020-7403. DOI: 10.1016/0020-7403(95)00105-0 (cit. on pp. 8, 91).

[24] H. Chen and D. Winterbone. "A method to predict performance of vaneless radial turbine under steady and unsteady flow conditions". In: *Turbocharging and Turbochargers*. Institution of Mechanical Engineers. 1990, pp. 13–22 (cit. on pp. 8, 20, 21, 97).

[25] J. R. Serrano, F. J. Arnau, V. Dolz, A. Tiseira, and C. Cervelló. "A model of turbocharger radial turbines appropriate to be used in zero- and one-dimensional gas dynamics codes for internal combustion engines modelling". *Energy Conversion and Management* 49(12) (2008), pp. 3729 –3745. ISSN: 0196-8904. DOI: 10.1016/j.enconman.2008.06.031 (cit. on pp. 8, 30, 31, 96, 99, 100, 134).

[26] J. R. Serrano, F. J. Arnau, R. Novella, and M. Á. Reyes-Belmonte. "A Procedure to Achieve 1D Predictive Modeling of Turbochargers under Hot and Pulsating Flow Conditions at the Turbine Inlet". *SAE Technical Paper* 2014-01-1080 (2014), 13pp. DOI: 10.4271/2014-01-1080 (cit. on p. 8).

[27] F. Payri, P. Olmeda, F. J. Arnau, A. Dombrovsky, and L. Smith. "External heat losses in small turbochargers: Model and experiments". *Energy* 71 (2014), pp. 534 –546. ISSN: 0360-5442. DOI: 10.1016/j.energy.2014.04.096 (cit. on pp. 8, 20, 27, 92, 131).

[28] J. R. Serrano, P. Olmeda, F. J. Arnau, A. Dombrovsky, and L. Smith. "Turbocharger heat transfer and mechanical losses influence in predicting engines performance by using one-dimensional simulation codes". *Energy* 86 (2015), pp. 204 –218. DOI: 10.1016/j.energy.2015.03.130 (cit. on pp. 8, 27, 92, 131).

[29] A. Gil, A. Tiseira, L. M. García-Cuevas, T. Rodríguez Usaquén, and G. Mijotte. "Fast three-dimensional heat transfer model for computing internal temperatures in the bearing housing of automotive turbochargers". *International Journal of Engine Research* (2018). DOI: 10.1177/1468087418804949 (cit. on pp. 8, 27, 92, 131).

[30] J. R. Serrano, F. J. Arnau, L. M. García-Cuevas, A. Dombrovsky, and H. Tartoussi. "Development and validation of a radial turbine efficiency and mass flow model at design and off-design conditions". *Energy Conversion and Management* 128 (2016), pp. 281 –293. ISSN: 0196-8904. DOI: 10.1016/j.enconman.2016.09.032 (cit. on pp. 8, 26–29, 96, 98, 131, 134, 135, 139, 157, 178).

[31] J. Galindo, R. Navarro, L. M. García-Cuevas, D. Tarí, H. Tartoussi, and S. Guilain. "A zonal approach for estimating pressure ratio at compressor extreme off-design conditions". *International Journal of Engine Research* 20(4) (2019), pp. 393–404. DOI: 10.1177/1468087418754899 (cit. on pp. 8, 92, 131).

[32] J. R. Serrano, P. Olmeda, A. Tiseira, L. M. García-Cuevas, and A. Lefebvre. "Theoretical and experimental study of mechanical losses in automotive turbochargers". *Energy* 55(0) (2013), pp. 888 –898. ISSN: 0360-5442. DOI: 10.1016/j.energy.2013.04.042 (cit. on pp. 8, 21, 92, 131).

[33] J. Galindo, H. Climent, A. Tiseira, and L. M. García-Cuevas. "Effect of the numerical scheme resolution on quasi-2D simulation of an automotive radial turbine under highly pulsating flow". *Journal of Computational and Applied Mathematics* 291 (Jan. 2016), pp. 112–126. DOI: 10.1016/j.cam.2015.02.025 (cit. on pp. 8, 131).

[34] J. Serrano, F. Arnau, L. Garcéa-Cuevas, and V. Samala. "Development of flow oriented model for extrapolation and interpolation off-design performance of twin-entry and dual volute radial-inflow turbines working under different flow admission conditions". *Energy* (). Under review (cit. on pp. 8, 92, 99, 100, 177).

[35] A. W. Costall, R. M. McDavid, R. F. Martínez-Botas, and N. C. Baines. "Pulse performance modelling of a twin-entry turbocharger turbine under full unequal admission". In: *Proceedings of ASME Turbo Expo 2009*. 2009. ASME, 2009. DOI: 10.1115/1.4000566 (cit. on pp. 8, 20, 130).

[36] A. W. Costall, R. M. McDavid, R. F. Martinez-Botas, and N. C. Baines. "Pulse Performance Modeling of a Twin Entry Turbocharger Turbine Under Full and Unequal Admission". *Journal of Turbomachinery* 133(2) (2011), p. 021005. ISSN: 0889504X. DOI: 10.1115/1.4000566 (cit. on pp. 8, 91).

[37]   A. Romagnoli, C. D. Copeland, R. Martinez-Botas, M. Seiler, S. Rajoo, and A. Costall. "Comparison Between the Steady Performance of Double-Entry and Twin-Entry Turbocharger Turbines". *Journal of Turbomachinery* 135(1) (2013). ISSN: 0889-504X. DOI: 10.1115/1.4006566 (cit. on pp. 8, 56).

[38]   M. S. Chiong, S. Rajoo, R. F. Martinez-Botas, and A. W. Costall. "Engine turbocharger performance prediction: One-dimensional modeling of a twin entry turbine". *Energy Conversion and Management* 57 (2012), pp. 68–78. ISSN: 01968904. DOI: 10.1016/j.enconman.2011.12.001 (cit. on pp. 8, 91).

[39]   M. S. Chiong, S. Rajoo, A. Romagnoli, A. W. Costall, and R. F. Martinez-Botas. "Assessment of Partial-Admission Characteristics in Twin-Entry Turbine Pulse Performance Modelling". In: *Volume 2C: Turbomachinery*. 2015, V02CT42A022. ISBN: 978-0-7918-5665-9. DOI: 10.1115/GT2015-42687 (cit. on pp. 8, 24, 92).

[40]   M. S. Chiong, S. Rajoo, A. Romagnoli, A. W. Costall, and R. F. Martinez-Botas. "One-dimensional pulse-flow modeling of a twin-scroll turbine". *Energy* 115 (2016), pp. 1291–1304. ISSN: 03605442. DOI: 10.1016/j.energy.2016.09.041 (cit. on p. 8).

[41]   M. S. Chiong, M. A. Abas, F. X. Tan, S. Rajoo, R. Martinez-Botas, Y. Fujita, T. Yokoyama, S. Ibaraki, and M. Ebisu. "Steady-State, Transient and WLTC Drive-Cycle Experimental Performance Comparison between Single-Scroll and Twin-Scroll Turbocharger Turbine". In: *WCX SAE World Congress Experience*. SAE International, 2019. DOI: https://doi.org/10.4271/2019-01-0327 (cit. on pp. 8, 56).

[42]   S. Rajoo, A. Romagnoli, and R. F. Martinez-Botas. "Unsteady performance analysis of a twin-entry variable geometry turbocharger turbine". *Energy* 38(1) (2012), pp. 176–189. ISSN: 03605442. DOI: 10.1016/j.energy.2011.12.017 (cit. on pp. 8, 57, 91).

[43]   S. Rajoo and R. F. Martinez-Botas. "Variable Geometry Mixed Flow Turbine for Turbochargers: An Experimental Study". *International Journal of Fluid Machinery and Systems* 1(1) (2008), pp. 155–168. ISSN: 1882-9554. DOI: 10.5293/IJFMS.2008.1.1.155 (cit. on pp. 8, 91).

[44]   C. D. Copeland, R. F. Martinez-Botas, and M. Seiler. "Comparison Between Steady and Unsteady Double-Entry Turbine Performance Using the Quasi-Steady Assumption". *Journal of Turbomachinery* 133(3) (2011), p. 031001. ISSN: 0889504X. DOI: 10.1115/1.4000580 (cit. on p. 8).

[45]   C. D. Copeland, R. F. Martinez-Botas, and M. Seiler. "Unsteady Performance of a Double Entry Turbocharger Turbine With a Comparison to Steady Flow Conditions". *Journal of Turbomachinery* 134(2) (2012), p. 021022. ISSN: 0889504X. DOI: 10.1115/1.4003171 (cit. on pp. 8, 91).

[46]   F. Payri. "Predicción de las actuaciones de los grupos de sobrealimentación para motores diesel de automoción". PhD book. Universitat Politècnica de València, 1973 (cit. on p. 9).

[47] J. R. Serrano. "Análisis y modelado de transitorios de carga en MEC turboalimentados". PhD book. Universitat Politècnica de València, 1999 (cit. on p. 9).

[48] C. Cervelló. "Contribución a la Caracterización Experimental y al Modelado de Turbinas de Geometría Variable en Grupos de Sobrealimentación". PhD book. Universitat Politècnica de València, 2005 (cit. on p. 9).

[49] A. Tiseira. "Caracterización experimental y modelado de bombeo en compresores centrífugos de sobrealimentación". PhD book. Universitat Politècnica de València, 2008 (cit. on p. 9).

[50] P. Fajardo. "Methodology for the Numerical Characterization of a Radial Turbine under Steady and Pulsating Flow". PhD book. Universitat Politècnica de València, 2012 (cit. on p. 9).

[51] R. Navarro. "A numerical approach for predicting flow-induced acoustics at near-stall conditions in an automotive turbocharger compressor". PhD book. Universitat Politècnica de València, 2014 (cit. on p. 9).

[52] M. Ángel Reyes-Belmonte. "Contribution to the Experimental Characterization and 1-D Modelling of Turbochargers for IC Engines". PhD book. Universitat Politècnica de València, 2013 (cit. on p. 9).

[53] L. M. G.-C. González. "Experiments and Modelling of Automotive Turbochargers under Unsteady Conditions". PhD book. Universitat Politècnica de València, 2014 (cit. on pp. 9–11, 176–178).

[54] A. Dombrovsky. "Synthesis of the 1D modelling of turbochargers and its effects on engine performance prediction". PhD book. Universitat Politècnica de València, 2016 (cit. on p. 9).

[55] L. Inhestern. "Measurement, Simulation, and 1D-Modeling of Turbocharger Radial Turbines at Design and Extreme Off-Design Conditions". PhD book. Universitat Politècnica de València, 2014 (cit. on p. 9).

[56] D. Tarí. "Effect of inlet configuration on the performance and durability of an automotive turbocharger compressor". PhD book. Universitat Politècnica de València, 2018 (cit. on p. 9).

[57] M. Hernández. "A non-linear quasi-3D model for air management modelling in engines". PhD book. Universitat Politècnica de València, 2018 (cit. on p. 9).

[58] J. García-Tíscar. "Experiments on Turbocharger Compressor Acoustics". PhD book. Universitat Politècnica de València, 2017. URL: http://hdl.handle.net/10251/79552 (cit. on p. 9).

[59] V. Samala. "Experimental characterization and mean line modelling of twin-entry and dual-volute turbines working under steady with different flow admission conditions". PhD book. Universitat Politècnica de València, 2020 (cit. on p. 9).

[60] U. Kesgin. "Effect of turbocharging system on the performance of a natural gas engine". *Energy Conversion and Management* 46(1) (2005), pp. 11–32. ISSN: 0196-8904. DOI: 10.1016/j.enconman.2004.02.006 (cit. on p. 20).

[61] H. Tang, A. Pennycott, S. Akehurst, and C. J. Brace. "A review of the application of variable geometry turbines to the downsized gasoline engine". *International Journal of Engine Research* 16(6) (2015), pp. 810–825. DOI: 10.1177/1468087414552289 (cit. on p. 20).

[62] A. Pesiridis. "The application of active control for turbocharger turbines". *International Journal of Engine Research* 13(4) (2012), pp. 385–398. DOI: 10.1177/1468087411435205 (cit. on p. 20).

[63] A. Romagnoli and R. Martinez-Botas. "Performance prediction of a nozzled and nozzleless mixed-flow turbine in steady conditions". *International Journal of Mechanical Sciences* 53(8) (2011), pp. 557–574. ISSN: 0020-7403. DOI: 10.1016/j.ijmecsci.2011.05.003 (cit. on p. 20).

[64] F. Payri, J. R. Serrano, P. Fajardo, M. A. Reyes-Belmonte, and R. Gozalbo-Belles. "A physically based methodology to extrapolate performance maps of radial turbines". *Energy Conversion and Management* 55(0) (2012), pp. 149–163. ISSN: 0196-8904. DOI: 10.1016/j.enconman.2011.11.003 (cit. on p. 20).

[65] V. De Bellis and S. Marelli. "One-dimensional simulations and experimental analysis of a wastegated turbine for automotive engines under unsteady flow conditions". *Proceedings of the Institution of Mechanical Engineers, Part D: Journal of Automobile Engineering* 229(13) (2015), pp. 1801–1816. DOI: 10.1177/0954407015571672 (cit. on p. 20).

[66] D. Bohn, N. Moritz, and M. Wolff. "Conjugate Flow and Heat Transfer Investigation of a Turbo Charger: Part II — Experimental Results". *ASME Conference Proceedings* 2003(3686) (2003), pp. 723–729. DOI: 10.1115/GT2003-38449 (cit. on p. 20).

[67] J. R. Serrano, P. Olmeda, A. Páez, and F. Vidal. "An experimental procedure to determine heat transfer properties of turbochargers". *Measurement Science and Technology* 21(3) (2010), p. 035109. DOI: 10.1088/0957-0233/21/3/035109 (cit. on pp. 20, 34, 95).

[68] J. Serrano, P. Olmeda, F. Arnau, and A. Dombrovsky. "General Procedure for the Determination of Heat Transfer Properties in Small Automotive Turbochargers". *SAE International Journal of Engines* 8(1) (2014), pp. 2014-01-2857. ISSN: 1946-3944. DOI: 10.4271/2014-01-2857 (cit. on p. 20).

[69] J. R. Serrano, P. Olmeda, F. J. Arnau, and V. Samala. "A holistic methodology to correct heat transfer and bearing friction losses from hot turbocharger maps in order to obtain adiabatic efficiency of the turbomachinery". *International Journal of Engine Research* (2019), p. 146808741983419. ISSN: 1468-0874. DOI: 10.1177/1468087419834194 (cit. on p. 20).

[70] R. Burke, C. Copeland, T. Duda, and M. Reyes-Belmote. "Lumped capacitance and three-dimensional computational fluid dynamics conjugate heat transfer modeling of an automotive turbocharger". *Journal of Engineering for Gas Turbines and Power* 138(9) (2016). DOI: 10.1115/1.4032663 (cit. on p. 20).

[71] X. Gao, B. Savic, and R. Baar. "A numerical procedure to model heat transfer in radial turbines for automotive engines". *Applied Thermal Engineering* (2019). ISSN: 13594311. DOI: 10.1016/j.applthermaleng.2019.03.014 (cit. on p. 20).

[72] P. Olmeda, V. Dolz, F. J. Arnau, and M. A. Reyes-Belmonte. "Determination of heat flows inside turbochargers by means of a one dimensional lumped model". *Mathematical and Computer Modelling* 57(7-8) (2013), pp. 1847 –1852. ISSN: 0895-7177. DOI: 10.1016/j.mcm.2011.11.078 (cit. on p. 21).

[73] J. R. Serrano, P. Olmeda, F. J. Arnau, M. A. Reyes-Belmonte, and A. Lefebvre. "Importance of Heat Transfer Phenomena in Small Turbochargers for Passenger Car Applications". *SAE Int. J. Engines* 6(2) (2 2013), pp. 716 –728. DOI: 10.4271/2013-01-0576 (cit. on p. 21).

[74] S. Marelli, G. Marmorato, and M. Capobianco. "Evaluation of heat transfer effects in small turbochargers by theoretical model and its experimental validation". *Energy* 112 (2016). ISSN: 03605442. DOI: 10.1016/j.energy.2016.06.067 (cit. on p. 21).

[75] A. Romagnoli, A. Manivannan, S. Rajoo, M. S. Chiong, A. Feneley, A. Pesiridis, and R. F. Martinez-Botas. *A review of heat transfer in turbochargers.* 2017. DOI: 10.1016/j.rser.2017.04.119 (cit. on p. 21).

[76] H. Aghaali, H.-E. Ångström, and J. R. Serrano. "Evaluation of different heat transfer conditions on an automotive turbocharger". *International Journal of Engine Research* 16(2) (2015), pp. 137–151. DOI: 10.1177/1468087414524755 (cit. on p. 21).

[77] S. Marelli, S. Gandolfi, and M. Capobianco. *Experimental and Numerical Analysis of Mechanical Friction Losses in Automotive Turbochargers.* SAE Technical Paper 2016-01-1026. SAE International, 2016. DOI: 10.4271/2016-01-1026 (cit. on p. 21).

[78] J. R. Serrano, P. Olmeda, A. Tiseira, L. M. García-Cuevas, and A. Lefebvre. "Importance of Mechanical Losses Modeling in the Performance Prediction of Radial Turbochargers under Pulsating Flow Conditions". *SAE Int. J. Engines* 6(2) (2 2013), pp. 729 –738. DOI: 10.4271/2013-01-0577 (cit. on p. 21).

[79] J. Galindo, P. Fajardo, R. Navarro, and L. M. García-Cuevas. "Characterization of a radial turbocharger turbine in pulsating flow by means of CFD and its application to engine modeling". *Applied Energy* 103(0) (2013), pp. 116 –127. ISSN: 0306-2619. DOI: 10.1016/j.apenergy.2012.09.013 (cit. on pp. 21, 56, 90, 130).

[80] J. Galindo, S. Hoyas, P. Fajardo, and R. Navarro. "Set-up analysis and optimization of CFD simulations for radial turbines". *Engineering Applications of Computational Fluid Mechanics* 7(4) (2013), pp. 441–460. DOI: 10.1080/19942060.2013.11015484 (cit. on p. 21).

[81] I Hakeem, C.-C. Su, A. Costall, and R. F. Martínez-Botas. "Effect of volute geometry on the steady and unsteady performance of mixed-flow turbines". In: *Proceedings of the Institution of Mechanical Engineers Part A-Journal of Power and Energy*. Vol. 221. 2007, pp. 535–550. DOI: 10.1243/09576509JPE314 (cit. on p. 21).

[82] X. Hu. "An advanced turbocharger model for the internal combustion engine". PhD book. Purdue University, 2000 (cit. on p. 21).

[83] A. King. "A turbocharger unsteady performance model for the GT-Power internal combustion engine simulation". PhD book. Purdue University, 2002 (cit. on p. 21).

[84] A. Feneley, A. Pesiridis, and H. Chen. "A one-dimensional gas dynamics code for turbocharger turbine pulsating flow performance modelling". In: *Proceedings of the ASME Turbo Expo*. Vol. 8. American Society of Mechanical Engineers (ASME), 2017. ISBN: 9780791850954. DOI: 10.1115/GT2017-64743 (cit. on p. 21).

[85] S. Rajoo and M.-B. Ricardo. "Variable Geometry Mixed Flow Turbine for Turbochargers: An Experimental Study". *International Journal of Fluid Machinery and Systems* 1(1) (Oct. 2008), pp. 155–168. DOI: 10.5293/IJFMS.2008.1.1.155 (cit. on pp. 21, 57).

[86] K. G. Hohenberg, P. J. Newton, R. F. Martinez-Botas, M. Halamek, K. Maeda, and J. Bouilly. "Development and Experimental Validation of a Low Order Tur-bine Model Under Highly Pulsating Flow". In: *Turbo Expo: Power for Land, Sea, and Air, Volume 2C: Turbomachinery*. ASME. June 2017, V02CT44A023. DOI: 10.1115/GT2017-63983 (cit. on pp. 21, 57, 90, 130).

[87] J. R. Serrano, A. Tiseira, L. M. García-Cuevas, L. Inhestern, and H. Tartoussi. "Radial turbine performance measurement under extreme off-design conditions". *Energy* 125 (2017), pp. 72–84. DOI: 10.1016/j.energy.2017.02.118 (cit. on pp. 21, 57, 90, 135).

[88] A. J. Torregrosa, A. Broatch, R. Navarro, and J. García-Tíscar. "Acoustic cha-racterization of automotive turbocompressors". *International Journal of Engine Research* 16(1) (2015), pp. 31–37. DOI: 10.1177/1468087414562866 (cit. on pp. 21, 57, 90, 130).

[89] O. Leufvén and L. Eriksson. "Measurement, analysis and modeling of centrifugal compressor flow for low pressure ratios". *International Journal of Engine Research* 17(2) (2016), pp. 153–168. DOI: 10.1177/1468087414562456 (cit. on p. 21).

[90] J. Galindo, A. Tiseira, R. Navarro, D. Tarí, and C. Meano. "Effect of the inlet geometry on performance, surge margin and noise emission of an automotive turbocharger compressor". *Applied Thermal Engineering* 110 (2017), pp. 875–882. DOI: 10.1016/j.applthermaleng.2016.08.099 (cit. on pp. 21, 130).

[91]   J. Galindo, A. Tiseira, P. Fajardo, and L. M. García-Cuevas. "Development and validation of a radial variable geometry turbine model for transient pulsating flow applications". *Energy Conversion and Management* 85 (2014), pp. 190–203. ISSN: 0196-8904. DOI: `10.1016/j.enconman.2014.05.072` (cit. on pp. 21, 25, 131).

[92]   Z. Ding, W. Zhuge, Y. Zhang, H. Chen, and R. Martinez-Botas. "Investigation on pulsating flow effect of a turbocharger turbine". In: *American Society of Mechanical Engineers, Fluids Engineering Division (Publication) FEDSM*. Vol. 1A-2017. 2017. ISBN: 9780791858042. DOI: `10.1115/FEDSM2017-69186` (cit. on p. 21).

[93]   Z. Ding, W. Zhuge, Y. Zhang, H. Chen, R. Martinez-Botas, and M. Yang. "A one-dimensional unsteady performance model for turbocharger turbines". *Energy* (2017). ISSN: 0360-5442. DOI: `10.1016/j.energy.2017.04.154` (cit. on pp. 21, 130, 131).

[94]   J. Galindo, J. R. Serrano, F. J. Arnau, and P. Piqueras. "Description of a Semi-Independent Time Discretization Methodology for a One-Dimensional Gas Dynamics Model". *Journal of Engineering for Gas Turbines and Power* 131(3) (2009), p. 034504. ISSN: 07424795. DOI: `10.1115/1.2983015` (cit. on pp. 23, 133).

[95]   S. K. Godunov. "A Difference Scheme for Numerical Solution of Discontinuous Solution of Hydrodynamic Equations". *Matematicheskii Sbornik* 47 (1959), pp. 271–306 (cit. on pp. 23, 178).

[96]   B. van Leer. "Towards the Ultimate Conservation Difference Scheme. II. Monotonicity and Conservation Combined in a Second-Order Scheme". *Journal of Computational Physics* 14 (Mar. 1974), p. 361. DOI: `10.1016/0021-9991(74)90019-9` (cit. on p. 23).

[97]   R. Courant, K. Friedrichs, and H. Lewy. "Über die partiellen Differenzengleichungen der mathematischen Physik". *Mathematische Annalen* 100(1) (Dec. 1928), pp. 32–74. ISSN: 0025-5831. DOI: `10.1007/bf01448839` (cit. on p. 24).

[98]   J. R. Serrano, F. J. Arnau, L. M. García-Cuevas, and L. B. Inhestern. "An innovative losses model for efficiency map fitting of vaneless and variable vaned radial turbines extrapolating towards extreme off-design conditions". *Energy* 180 (2019), 626–639. ISSN: 0360-5442. DOI: `10.1016/j.energy.2019.05.062` (cit. on pp. 26, 157).

[99]   *Supercharger Testing Standard*. SAE J1723. Society of Automotive Engineers, 1995 (cit. on p. 33).

[100]   *Turbocharger gas stand test code*. SAE J1826. Society of Automotive Engineers, 1995 (cit. on p. 33).

[101]   G Pinero, L Vergara, J. M. Desantes, and A Broatch. "Estimation of velocity fluctuation in internal combustion engine exhaust systems through beamforming techniques". *Measurement Science and Technology* 11(11) (2000), p. 1585. DOI: `10.1088/0957-0233/11/11/307` (cit. on pp. 36, 39, 64, 133, 177).

[102]   P. Welch. "The Use of Fast Fourier Transform for the Estimation of Power Spectra: A Method Based on Time Averaging Over Short, Modified Periodograms". *IEEE Transactions on Audio Electroacoustics* 15 (2 1967), pp. 70–73. ISSN: 0018-9278. DOI: 10.1109/TAU.1967.1161901 (cit. on pp. 39, 106, 136).

[103]   F. Harris. "On the use of windows for harmonic analysis with the discrete Fourier transform". In: *Proceedings of the IEEE*. Vol. 66. IEEE, 1978, pp. 51–83. DOI: 10.1109/PROC.1978.10837 (cit. on pp. 39, 106).

[104]   R. Hogg. "Life beyond euro VI". *automotiveworld* (2014). URL: http://www.automotiveworld.com/megatrends-articles/life-beyond-euro-vi/ (cit. on pp. 56, 90).

[105]   M. Knopf. "How low can we go? Downsizing the internal combustion engine". *Ingenia* (2011). URL: https://www.ingenia.org.uk/Ingenia/Articles/11612c67-32db-477a-af8f-38cef44ee7b2 (cit. on pp. 56, 90).

[106]   E. Watel, A. Pagot, P. Pacaud, and J.-C. Schmitt. "Matching and Evaluating Methods for Euro 6 and Efficient Two-stage Turbocharging Diesel Engine". In: *SAE 2010 World Congress and Exhibition*. SAE international. 2010. DOI: 10.4271/2010-01-1229 (cit. on p. 56).

[107]   F. Pischinger and A. Wuensche. "The characteristic behaviour of radial turbines and its influence on the turbocharging process". In: *Proceedings of the CIMAC Conference, Tokyo, Japan*. 1977, pp. 545–568 (cit. on p. 56).

[108]   A. Torregrosa, J. Serrano, J. Dopazo, and S. Soltani. "Experiments on wave transmission and reflection by turbochargers in engine operating conditions". *SAE Technical Papers* 2006(01:0022) (2006). DOI: 10.4271/2006-01-0022 (cit. on pp. 57, 90).

[109]   A. Costall, V. Cheong, H. Flora, A. Munasinghe, R. Ivanov, R. W. Kruiswyk, and J. R. McDonald. "Development of a Novel Transient-Pulsating Flow Rig for Engine Air System Research using GT-SUITE". In: *European GT Conference*. 2018 (cit. on pp. 57, 180).

[110]   C. Arcoumanis, I. Hakeem, L. Khezzar, R. F. Martinez-Botas, and N. C. Baines. "Performance of a Mixed Flow Turbocharger Turbine Under Pulsating Flow Conditions". In: *Volume 2: Aircraft Engine; Marine; Microturbines and Small Turbomachinery*. ASME, 1995, V002T04A011. ISBN: 978-0-7918-7879-8. DOI: 10.1115/95-GT-210 (cit. on p. 57).

[111]   C. Copeland, R. Martinez-Botas, and M. Seiler. "Unsteady performance of a double entry turbocharger turbine with a comparison to steady flow conditions". In: *Proceedings of the ASME Turbo Expo*. Proceedings of the ASME Turbo Expo. ASME, 2008. DOI: 10.1115/GT2008-508279 (cit. on p. 57).

[112]   M. Yang, K. Deng, R. Martines-Botas, and W. Zhuge. "An investigation on unsteadiness of a mixed-flow turbine under pulsating conditions". *Energy Conversion and Management* 110 (2016), pp. 51–58. ISSN: 01968904. DOI: 10.1016/j.enconman.2015.12.007 (cit. on p. 57).

[113] C. D. Copeland, R. F. Martínez-Botas, and M Seiler. "Comparison between steady and unsteady double-entry turbine performance using the quasi-steady assumption". *Journal of Turbomachinery: Transactions of the ASME* 133 (3 2010). DOI: 10.1115/1.4000580 (cit. on p. 57).

[114] M. Yang, R. F. Martinez-Botas, S. Rajoo, T. Yokoyama, and S. Ibaraki. "Influence of Volute Cross-Sectional Shape of a Nozzleless Turbocharger Turbine Under Pulsating Flow Conditions". In: *Volume 2D: Turbomachinery*. ASME, 2014, V02DT42A025. ISBN: 978-0-7918-4563-9. DOI: 10.1115/GT2014-26150 (cit. on pp. 57, 91).

[115] M. Cerdoun and A. Ghenaiet. "Unsteady behaviour of a twin entry radial turbine under engine like inlet flow conditions". *Applied Thermal Engineering* 130 (2018), pp. 93–111. ISSN: 1359-4311. DOI: 10.1016/j.applthermaleng.2017.11.001 (cit. on pp. 57, 91).

[116] C. Cravero, D. De Domenico, and A. Ottonello. "Numerical Simulation of the Performance of a Twin Scroll Radial Turbine at Different Operating Conditions". *International Journal of Rotating Machinery* 5302145 (2019). ISSN: 15423034. DOI: 10.1155/2019/5302145 (cit. on p. 57).

[117] R. Kabral, Y. El Nemr, C. Ludwig, R. Mirlach, P. Koutsovasilis, A. Masrane, and M. Åbom. "Experimental acoustic characterization of automotive twin-scroll turbine". In: *12th European Conference on Turbomachinery Fluid Dynamics and Thermodynamics ETC*. 2017, p. 363 (cit. on pp. 57, 75, 92).

[118] N. Watson and M. Janota. "Turbocharging the internal combustion engine". *The Macmillan Press Ltd* (1982) (cit. on p. 90).

[119] D. Winterbone, B Nikpour, and H Frost. "A contribution to the understanding of turbocharger turbine performance in pulsating flow". *Inst Mech Eng Semin, paper C433011* (1991) (cit. on p. 90).

[120] D. Winterbone and R. Pearson. "Turbocharger turbine performance under unsteady flow-a review of experimental results and proposed models". *Inst Mech Eng Semin, paper C55403198* (1998) (cit. on pp. 90, 130).

[121] F. Piscaglia, A. Onorati, S. Marelli, and M. Capobianco. "A detailed one-dimensional model to predict the unsteady behavior of turbocharger turbines for internal combustion engine applications". *International Journal of Engine Research* 20(3) (2019), pp. 327–349. DOI: 10.1177/1468087417752525 (cit. on pp. 90, 130).

[122] N. Cappelaere, A. Dazin, and G. Bois. "An industrial experimental methodology for the unsteady characterization of automotive turbocharger". *16th International Symposium on Transport Phe- nomena and Dynamics of Rotating Machinery* (2016) (cit. on p. 91).

[123] N. Baines, A. Hajilouy-Benisi, and J. Yeo. "The pulse flow performance and modelling of radial inflow turbines". In: *5th International Conference on Turbochargers, Inst Mech Eng, London, Paper C48400694*. 1994, pp. 209–219 (cit. on p. 91).

[124] C. D. Copeland, R. F. Martinez-Botas, and M. Seiler. "Comparison Between Steady and Unsteady Double-Entry Turbine Performance Using the Quasi-Steady Assumption". *Journal of Turbomachinery* 133(3) (2011), p. 031001. ISSN: 0889504X. DOI: 10.1115/1.4000580 (cit. on p. 91).

[125] M. Yang, R. F. Martinez-Botas, S. Rajoo, T. Yokoyama, and S. Ibaraki. "An investigation of volute cross-sectional shape on turbocharger turbine under pulsating conditions in internal combustion engine". *Energy Conversion and Management* 105 (2015). ISSN: 01968904. DOI: 10.1016/j.enconman.2015.06.038 (cit. on p. 91).

[126] C. D. Copeland, P. J. Newton, R. F. Martinez-Botas, and M. Seiler. "The Effect of Unequal Admission on the Performance and Loss Generation in a Double-Entry Turbocharger Turbine". *Journal of Turbomachinery* 134(2) (2012), p. 021004. ISSN: 0889504X. DOI: 10.1115/1.4003226 (cit. on p. 91).

[127] J. Macek, Z. Zak, and O. Vitek. "Physical Model of a Twin-scroll Turbine with Unsteady Flow". *SAE Technical Papers* (2015). DOI: 10.4271/2015-01-1718 (cit. on p. 91).

[128] M. S. Chiong, S. Rajoo, A. Romagnoli, A. W. Costall, and R. F. Martinez-Botas. "One-dimensional pulse-flow modeling of a twin-scroll turbine". *Energy* 115 (2016), pp. 1291–1304. ISSN: 03605442. DOI: 10.1016/j.energy.2016.09.041 (cit. on p. 92).

[129] J. Serrano, F. Arnau, L. García-Cuevas, V. Samala, and L. Smith. "Experimental approach for the characterization and performance analysis of twin entry radial-inflow turbines in a gas stand and with different flow admission conditions". *Applied Thermal Engineering* 159 (2019), p. 113737. DOI: 10.1016/j.applthermaleng.2019.113737 (cit. on pp. 92, 95, 98, 102, 115).

[130] R. Baar, C. Biet, V. Boxberger, H. Mai, and R. Zimmermann. "New Evaluation of Turbocharger Components based on Turbine Outlet Temperature Measurements in Adiabatic Conditions". In: *15th International Symposium on Transport Phenomena and Dynamics of Rotating Machinery*. 2014. 2014 (cit. on p. 95).

[131] R Zimmermann, R Baar, and C Biet. "Determination of the isentropic turbine efficiency due to adiabatic measurements and the validation of the conditions via a new criterion". *Proceedings of the Institution of Mechanical Engineers, art C: Journal of Mechanical Engineering Science* 232(24) (2018), pp. 4485–4494. DOI: 10.1177/0954406216670683 (cit. on p. 95).

[132] J. R. Serrano, F. J. Arnau, P. Fajardo, and M. A. Reyes-Belmonte. "Contribution to the Modeling and Understanding of Cold Pulsating Flow Influence in the Efficiency of Small Radial Turbines for Turbochargers". *Journal of Engineering for Gas Turbines and Power* 134(10) (2012) (cit. on pp. 96, 98).

[133] H. Chen, I. Hakeem, and R. F. Martínez-Botas. "Modelling of a turbocharger turbine under pulsating inlet conditions". In: *Proceedings of the Institution of Mechanical Engineers, Part A: Journal of Power and Energy*. Institution of Mechanical Engineers. Oct. 1996, pp. 397–408 (cit. on p. 97).

[134] H. Rämmal and M. Åbom. "Acoustics of Turbochargers". In: *SAE 2007 Noise and Vibration Conference and Exhibition*. SAE International, 2007. DOI: 10.4271/2007-01-2205 (cit. on p. 130).

[135] A. Broatch, J. Galindo, R. Navarro, and J. García-Tíscar. "Methodology for experimental validation of a CFD model for predicting noise generation in centrifugal compressors". *International Journal of Heat and Fluid Flow* 50 (2014), 134–144. DOI: 10.1016/j.ijheatfluidflow.2014.06.006 (cit. on p. 130).

[136] A. Broatch, X. Margot, J. García-Tíscar, and F. Roig. "Impact of simple surge-enhancing inlet geometries on the acoustic behavior of a turbocompressor". *International Journal of Engine Research* (2018). DOI: 10.1177/1468087418784125 (cit. on p. 130).

[137] S. Sharma, A. Broatch, J. García-Tíscar, J. M. Allport, and A. K. Nickson. "Acoustic characteristics of a ported shroud turbocompressor operating at design conditions". *International Journal of Engine Research* (2018). DOI: 10.1177/1468087418814635 (cit. on p. 130).

[138] K. Peat, A. Torregrosa, A. Broatch, and T. Fernández. "An investigation into the passive acoustic effect of the turbine in an automotive turbocharger". *Journal of Sound and Vibration* 295 (1-2 Aug. 2006), pp. 60–75. DOI: 10.1016/j.jsv.2005.11.033 (cit. on p. 130).

[139] A. Torregrosa, J. Galindo, J. R. Serrano, and A. Tiseira. "A Procedure for the Unsteady Characterization of Turbochargers in Reciprocating Internal Combustion Engines". In: *Fluid Machinery and Fluid Mechanics*. Ed. by J. Xu, Y. Wu, Y. Zhang, and J. Zhang. Berlin, Heidelberg: Springer Berlin Heidelberg, 2009, pp. 72–79. DOI: 10.1007/978-3-540-89749-1_10 (cit. on p. 130).

[140] R. Kabral, Y. A. El Nemr, C. Ludwig, R. Mirlach, P. Koutsovasilis, A. Masrane, and M. Åbom. "Experimental acoustic characterization of automotive twin-scroll turbine". In: *12th European Conference on Turbomachinery Fluid dynamics & Thermodynamics*. European Turbomachinery Society. DOI: 10.29008/ETC2017-363 (cit. on p. 130).

[141] A. Broatch, J. Galindo, R. Navarro, J. García-Tíscar, A. Daglish, and R. K. Sharma. "Simulations and measurements of automotive turbocharger compressor whoosh noise". *Engineering Applications of Computational Fluid Mechanics* 9(1) (2015). DOI: 10.1080/19942060.2015.1004788 (cit. on p. 130).

[142]   A. Broatch, J. Galindo, R. Navarro, and J. García-Tíscar. "Numerical and experimental analysis of automotive turbocharger compressor aeroacoustics at different operating conditions". *International Journal of Heat and Fluid Flow* 61 (2016), pp. 245–255. DOI: 10.1016/j.ijheatfluidflow.2016.04.003 (cit. on p. 130).

[143]   A. Marsan and S. Moreau. "Analysis of the flow structure in a radial turbine". In: *11th European Conference on Turbomachinery Fluid Dynamics and Thermodynamics, ETC 2015*. Madrid, Spain, Mar. 2015 (cit. on p. 130).

[144]   F. J. Wallace and J. Adgey. "Paper 1: Theoretical Assessment of the Non-Steady Flow Performance of Inward Radial Flow Turbines". *Proceedings of the Institution of Mechanical Engineers, Conference Proceedings* 182(8) (1967), pp. 22–36. DOI: 10.1243/PIME_CONF_1967_182_211_02 (cit. on p. 130).

[145]   M. S. Chiong, S. Rajoo, A. Romagnoli, and R. Martínez-Botas. "Unsteady performance prediction of a single entry mixed flow turbine using 1-D gas dynamic code extended with meanline model". In: *Proceedings of the ASME Turbo Expo*. Vol. 5. ASME. 2012, pp. 781–795. DOI: 10.1115/GT2012-69176 (cit. on p. 131).

[146]   C. Avola, C. Copeland, A. Romagnoli, R. Burke, and P. Dimitriou. "Attempt to correlate simulations and measurements of turbine performance under pulsating flows for automotive turbochargers". *Proceedings of the Institution of Mechanical Engineers, Part D: Journal of Automobile Engineering* 233(2) (2019), pp. 174–187. DOI: 10.1177/0954407017739123 (cit. on p. 131).

[147]   R. Veloso, Y. Elnemr, F. M. Reich, and S. Allam. "Simulation of Sound Transmission through Automotive Turbochargers". In: *7th International Styrian Noise, Vibration & Harshness Congress: The European Automotive Noise Conference*. SAE International, June 2012. DOI: 10.4271/2012-01-1560 (cit. on p. 131).

[148]   J. Galindo, A. Tiseira, R. Navarro, D. Tarí, H. Tartoussi, and S. Guilain. "Compressor Efficiency Extrapolation for 0D-1D Engine Simulations". In: *SAE 2016 World Congress and Exhibition*. SAE Technical Paper 2016-01-0554. SAE International, 2016. DOI: 10.4271/2016-01-0554 (cit. on p. 131).

[149]   Z. Zhu, S. Midlam-Mohler, and M. Canova. "Development of physics-based three-way catalytic converter model for real-time distributed temperature prediction using proper orthogonal decomposition and collocation". *International Journal of Engine Research* (2019). DOI: 10.1177/1468087419876127 (cit. on p. 150).

[150]   D. Gordon, C. Wouters, M. Wick, F. Xia, B. Lehrheuer, J. Andert, C. R. Koch, and S. Pischinger. "Development and experimental validation of a real-time capable field programmable gate array–based gas exchange model for negative valve overlap". *International Journal of Engine Research* (2019). DOI: 10.1177/1468087418788491 (cit. on p. 150).

www.ingramcontent.com/pod-product-compliance
Lightning Source LLC
LaVergne TN
LVHW040017200726
843493LV00005B/1292